GEOGRAPHIES OF RE

Geographies of Rhythm
Nature, Place, Mobilities and Bodies

Edited by
TIM EDENSOR
Manchester Metropolitan University, UK

Routledge
Taylor & Francis Group

LONDON AND NEW YORK

First published 2010 by Ashgate Publishing

2 Park Square, Milton Park, Abingdon, Oxon OX14 4RN
711 Third Avenue, New York, NY 10017, USA

Routledge is an imprint of the Taylor & Francis Group, an informa business

First issued in paperback 2016

British Library Cataloguing in Publication Data
Geographies of rhythm : nature, place, mobilities and
  bodies.
  1. Human beings--Effect of environment on. 2. Geographical
  perception. 3. Spatial behaviour.
  I. Edensor, Tim, 1957-
  304.2'3-dc22

Library of Congress Cataloging-in-Publication Data
Edensor, Tim, 1957-
  Geographies of rhythm : nature, place, mobilities and bodies / edited by Tim Edensor.
    p. cm.
  Includes bibliographical references and index.
  ISBN 978-0-7546-7662-1 (hardback) -- ISBN 978-0-7546-7663-8 (ebook)
  1. Rhythm. 2. Space and time. 3. Human geography. I. Title.

BH301.R5E34 2009
304.2'3--dc22

2009034968

ISBN 13: 978-0-7546-7662-1 (hbk)
ISBN 13: 978-1-138-27454-9 (pbk)

Contents

List of Figures

List of Contributors

Deirdre Conlon researches urban and global studies at the Institute for Liberal Arts and Interdisciplinary Studies, Emerson College, Boston. Currently focusing on contemporary Irish society, her research examines the social, cultural and political geographies of migration under conditions of global economic restructuring. She is editing a theme issue on 'Geographies of Waiting in Transnational Lives' for *Gender, Place, and Culture.*

Monica Degen teaches Cultural Sociology at Brunel University. She has published on topics relating to Barcelona's urban planning, the politics of public space, and the relationship between the senses and city life. She is author of *Sensing Cities* (Routledge, 2008) and co-author of *La Metaciudad: Barcelona* (Anthropos, 2008).

Dydia DeLyser is Associate Professor of Geography at Louisiana State University. Her book, *Ramona Memories: Tourism and the Shaping of Southern California* (University of Minnesota Press, 2005) won the AAG's Globe Book Award. Her current research focuses on issues of gender and mobilities among early women aviators.

Tim Edensor teaches Cultural Geography at Manchester Metropolitan University. He is the author of *Tourist at the Taj* (Routledge, 1998); *National Identity, Popular Culture and Everyday Life* (Berg 2002) and *Industrial Ruins: Space, Aesthetics and Materiality* (Berg, 2005). He is currently researching urban materialities and landscapes of illumination.

James Evans lectures in the Department of Geography at the University of Manchester. His research explores the politics of environmental governance, focusing on ecological planning, urban sustainability and questions of interdisciplinarity.

Rhys Evans operates a private consultancy, Integrate Consulting, which specialises in multi-method social research, policy analysis, evaluation and training in Scotland, the UK, and internationally. His research has focused upon community woodlands in Scotland and social capital and community capacity building in the Highlands and Islands. He is presently exploring the geographies of equine activities in the UK.

Alexandra Franklin is Research Manager and Senior Research Associate at the Research Centre for Business Relations Accountability, Sustainability and Society (BRASS), Cardiff University. Her research interests include sustainable communities, nature-society relations, people and place, public participation and rural development. She has authored several academic journal publications and policy reports for the Welsh Assembly, Scottish Executive, DEFRA and the Forestry Commission.

Tom Hall teaches anthropology and sociology at Cardiff University. He has carried out ethnographic research on different aspects of youth and street homelessness and is the author of *Better Times Than This: Youth Homelessness in Britain* (Pluto Press, 2003). He is currently working on a book on urban patrol and public space.

Shannon Hensley works in the Department of Geography at Exeter University. She works at the intersections of performance, public art, the body and the city. She also has a professional degree in architecture. Shannon spent her childhood in various parts of Indiana, Georgia, Louisiana and Kentucky and quickly realised the difference that place makes.

Richard Hornsey is Senior Lecturer in Cultural Studies at the University of the West of England, Bristol. His research focuses upon urban cultural history, the politics and poetics of everyday life and quotidian queer ethics. He is currently completing *The Spiv and the Architect: Space, Time and Male Homosexuality in Post-war London* (University of Minnesota Press, 2009).

Paola Jiron works at the Faculty of Architecture and Urbanism at the University of Chile. She has carried out extensive research, teaching and consultancy work in the areas of housing, urban quality of life, and urban mobility practices.

Owain Jones is a Research Fellow at the Countryside and Community Research Institute at the University of Gloucester. He has written widely on rurality, biodiversity, childhood, memory and non-representational theory, and is the co-author of *Tree Cultures: The Place of Trees and Trees in their Place* (Berg, 2002). He is currently researching tidal space and cultures.

Craig Meadows is currently researching insomnia and related cultural products for his PhD dissertation in Social and Political Thought at York University in Toronto, Canada. These interests arise from 20 years of chronic insomnia and from ten years of work in the culture industry as a buyer, programmer and reviewer prior to a return to academia.

Tracey Potts is Lecturer in Critical Theory and Cultural Studies at the University of Nottingham. Her research interests revolve around questions of taste, aesthetics and material culture. She is currently completing a book about kitsch and investigating landscapes of vernacular memory, links between creativity and waste disposal and popular strategies to minimise clutter of all sorts.

Justin Spinney is a Research Fellow for the Lifestyles, Values and the Environment (RESOLVE) institute at the University of Surrey UK, investigating how companies and individuals respond to issues of sustainability in the production and consumption of goods and services. He has researched the production and consumption of mobilities, embodiment, sensory geographies and ethnographic and video methods.

Filipa Matos Wunderlich is a Research Associate at the Bartlett School of Architecture and Planning where she completed her PhD on *Place-temporality in Urban Analysis and Design: Uncovering Everyday Urban Place-rhythms (at Fitzroy Square)*. She also holds a Diploma in Architecture, University of Porto, Portugal, and a Bachelor's degree in Music, ArtEZ Institute of Arts, in Zwolle, The Netherlands.

Chapter 1

Introduction: Thinking about Rhythm and Space

Tim Edensor

This edited collection aims to explore the fertile suggestions offered by Henri Lefebvre's *Rhythmanalysis*. First published in English in 2004, it has proved to be a stimulating resource for current thinking about timespace, place and everyday life. This volume features a diverse range of essays that have engaged with Lefebvre's ideas, developing some of his insights and going beyond his analysis. Thus, while *Rhythmanalysis* looms large throughout the book as a touchstone and point of embarkation, the ideas here are not constrained by Lefebvre's highly suggestive but brief volume. More specifically, the collection is concerned with investigating how rhythms shape human experience in timespace and pervade everyday life and place. In this introduction, through reviewing the critical and theoretical potential that rhythmanalysis offers and its intersection with current ideas in human geography and the social sciences, I introduce the chapters that follow.

Most obviously, rhythmanalysis, placed in the broader context of time-geography, can contribute to the development of the temporal understanding of place and space. Here, the cultural experience and social understandings of time must be conceived as dynamic, multiple and heterogeneous. Rather than 'a singular or uniform social time stretching across a uniform social space', May and Thrift contend that we need to be 'aware of various (and uneven) networks of time stretching in different and divergent directions across an uneven social field' (2001: 5) and Barbara Adam asserts that we need to explore the distinct formations of 'tempo, timing, duration, sequence and rhythm as the mutually implicating structures of time' (1998: 202). In this context, rhythmanalysis is particularly useful in investigating the patterning of a range of multiscalar temporalities – calendrical, diurnal and lunar, lifecycle, somatic and mechanical – whose rhythms provide an important constituent of the experience and organisation of social time.

The most obvious antecedent of rhythmanalyses is Hagerstrand's time-geography. Mels points out the failings of its diagrammatic schemes which suggest that space is somewhat empty, minimise the scale and breadth of routines, and present individual rhythms as rather unsensual and disembodied. However, despite not telling us much about how time-space is construed or experienced, time-geography demonstrates that individuals 'repeatedly couple and uncouple their paths with other people's paths, institutions, technologies and physical surroundings' (2004: 16) through which they become grounded in time-space and place. Rhythmanalysis can develop a fuller, richer analysis of these synchronic practices in space while also accounting for spatial qualities, sensations and intersubjective habits.

All chapters in this book implicitly or explicitly discuss the everyday, a realm within which regulatory processes pervade and are resisted or ignored. For rhythmanalysis is a useful tool with which to explore the everyday temporal structures and processes that (re)produce connections between individuals and the social. Lefebvre is explicit that there is no 'rhythm without repetition in time and space, without reprises, without returns, in short, without measure'. However, he is also insistent that 'there is no identical absolute repetition indefinitely ... there is always something new and unforeseen that introduces itself into the repetitive' (2004: 6). With this focus on multiple quotidian rhythms, we may identify how power is instantiated in unreflexive, normative practices but also side-stepped, resisted and supplemented by other dimensions of everyday experience. Barbara Adam also draws attention to how the 'when, how often, how long, in what order and at what speed' are governed by 'norms, habits and conventions' about temporality (1995: 66), a host of implicit, embedded and embodied forms of social knowing that regulate social life and space. In identifying the regulatory rhythmic conventions that shape the lives of individuals and groups, we can explore how some conform to dominant routines and timetables, while others reject such temporal structurings or become sidelined because they are thought to be out of time and step. The main focus of this introductory chapter, and a theme that resounds throughout this book, is the mix of social ordering and disordering through which spatio-temporal patterns are laid down. I explore the production of normative everyday rhythms and the ordering of timespace through state and capitalist processes shortly, and follow this with a discussion of how this is always only ever partial and susceptible to disordering by counter rhythms and arrhythmia. First, however, I briefly summarise the spatial, mobile, embodied and 'non-human' elements of rhythm which partly inform the organisation of the chapters in this book.

Placing rhythm: key themes

> You get the full essence of Harlem in an air shaft. You hear fights, you smell dinner, you hear people making love. You hear intimate gossip floating down. You hear the radio. An air shaft is one great big loudspeaker. You see your neighbor's laundry. You hear the janitor's dogs. The man upstairs' aerial falls down and breaks your window. You smell coffee ... An air shaft has got every contrast ... You hear people praying, fighting, snoring ... I tried to put all that in my Harlem Air Shaft. (Duke Ellington, quoted in Shapiro and Hentoff, 1955: 224-25)

This book starts from Lefebvre's premise that '(E)verywhere where there is interaction between a place, a time, and an expenditure of energy, there is *rhythm*' (2004: 15), identifying repetition of movements and action, the particular entanglements of linear and cyclical rhythms and phases of growth and decline. In identifying the spatio-temporal specificities of place, he further contends that 'every rhythm implies the relation of a time with space, a localised time, or if one wishes, a temporalised place' (1996: 230). We can identify the distinctive characteristics of place according to its 'polyrhythmic ensemble' (Crang, 2000), the particular ways in which changing rhythmic processes interweave to afford places a mixity of temporal events of varying regularity, as Duke Ellington captures both in his commentary and the rhythmically swinging composition of *Harlem Airshaft*. Such rhythms shape the diurnal, weekly and annual experience of place and influence the ongoing formation of its materiality. This perspective avoids the conception of place as static, for rhythms are essentially dynamic, part of the multiplicity of flows that emanate from, pass through and centre upon place, and contribute to its situated dynamics. Moreover, those rhythms that emerge from human practices are part of the continuous process 'of emplaced engagement with the material, sensory, social and cultural contexts in which we dwell' (Pink, 2007: 62).

Despite this fluidity and dynamism however, and the always immanent potential for disruption and destruction, many rhythms offer a consistency to place and landscape over time. For regardless of the ongoing becoming of life and place, regular routines and slower processes of change mesh with the relative brevity of the human lifespan to provide some sense of stability. Arjun Appadurai (1990) conceives globalisation as constituted by often disjunctive flows of people, commodities, information, ideas, technologies and finance. Consequently, we can conceive places as part of infinitely complex spatial networks (Massey, 1995) and not self-contained envelopes. Cities, for instance, are ceaselessly (re)constituted out of their connections, the 'twists and fluxes of interrelation' (Amin and Thrift, 2002: 30) through which 'multiple networked mobilities of capital, persons, objects, signs and information' (Urry, 2006: ix) are brought together to produce a particular, but ever-changing, complex mix of heterogeneous social interactions, materialities, mobilities, imaginaries and social effects. In fact, the ontological resilience of place 'depends upon the continuous flow of goods, people and capital outside and underneath its premises' (Kaika, 2005: 8) in consistent surges. For instance, the indefatigable maintenance of largely invisible key flows of water, electricity, gas and telephony is vital to secure the security and stability of the city. Rhythmanalysis can help explore notions that places are always in a process of becoming, seething with emergent properties, but usually stabilised by regular patterns of flow that possess particular rhythmic qualities whether steady, intermittent, volatile or surging. This implies that the spatial scale through which rhythms resound needs to be accounted for, for instance, how national and global rhythms increasingly pulse through place. In accounting for the constituent elements of rhythmic geographies which feature throughout this book, I now briefly discuss rhythms of people, bodies, mobilities and nature.

The rhythms of people

Allen (in Amin and Thrift, 2002: 17) points to 'the regular comings and goings of people about the city to the vast range of repetitive activities, sounds and even smells that punctuate life in the city and which give many of those who live there a sense of time and location'. Consider these routine, daily flows of people through space and place; the walking patterns of schoolchildren, the rush hour of commuters, the surge of shoppers, the throngs of evening clubbers, the rituals of housework, the lifestyles of students, the slow pace of unemployment, the timed compulsions of drug addicts and alcoholics, and the timetabled activities of tourists, to name but a few. Consider the 'openings and closing of shops, the flows of postal deliveries, bank deposits and coffee breaks' (Labelle, 2008: 192) as well as the schedules of public transport, pub hours and lighting up times, and the different rhythms of the day and night (see Sandhu, 2006), as well as seasonal and annual cycles, which bestow a temporal sense of place.

While regular, normative rhythms of place, often supported and promoted by officials and commercial enterprises, add to the knowing and feeling of place, they are frequently contested and disrupted by other inhabitants and passers-through. In this book, Tom Hall shows how the rhythms of the homeless clash with those desired by city managers and business, Craig Meadows reveals how insomniacs are out-of-synch with dominant diurnal beats, Justin Spinney and Richard Hornsey show how cyclists and pedestrians violate the rhythmic norms of vehicular traffic, and Deirdre Conlon shows how Nigerian asylum seekers lay down an offbeat to the rhythms of regeneration and consumerism. Place can thus be depicted, performed and sensed through its ensemble of normative and counter rhythms, as Lefebvre suggests. The obvious example of a train station reveals a very recognisable, though shifting polyrhythmy composed out of separate strands, with its periodic announcements, flows and surges of passengers, departing and arriving trains according (or not) to the timetable, the presence of newspaper sellers during rush hours, and the ongoing pulse of buying and selling in the retail outlets, as well as the interruptions, unexpected incidents and breakdowns. In this volume, Monica Degen highlights the regular but changing rhythms of an urban area undergoing regeneration, and Filipa Matos Wunderlich portrays the complex sensual, experiential, performative and everyday lineaments of the rhythms of a London square, focusing on the distinct sense of flow collectively produced and the particular soundscape produced by regular social activities and shared cultural practices.

Bodily rhythms

Lefebvre foregrounds the body, emphasising that the rhythmanalyst must draw on 'his (sic) breathing, the circulation of his blood, the beatings of his heart and the delivery of his speech as landmarks' (2004: 21), and recognise that rhythms are folded in and through the permeable body. The rhythmanalyst must take their own body – 'its respirations, pulses, circulations, assimilations ... durations and phases of durations' – as the measure of other rhythms. A disembodied appreciation of rhythms is impossible, for 'to listen to one's own body is necessary to 'appreciate external rhythms (ibid: 19). Yet as Simpson has pointed out, Lefebvre's focus too frequently turns to epistemological considerations and social disciplining of the body rather than actual embodied experience, 'the visceral, elusory nature' of the body (2008: 824) and its capacity to affect and be affected by a multitude of other rhythms. Lefebvre identifies the regulation of embodied rhythms through the notion of 'dressage' as a means to train the body to perform and condition it to accede to particular rhythms. However, the body also produces place as well as fitting in with it, and it may not keep in step or synchronise with regular beats. And while dressage may produce conformist rhythmic performance, it also has the potential to produce identity and scope for improvisation, as Shannon Hensley (this volume) shows in her account about how Cuban places and people associated with the performance of rumba, absorb rhythm into their bodies through prolonged practice so that it becomes a 'second nature' that masquerades as a racialised 'natural rhythm'.

This emphasis on the entangling rhythms that circulate in and outside the body also draws attention to the corporeal capacities to sense rhythm, sensations that organise the subjective and cultural experience of place. The usually unreflexive sensual and rhythmic attunement to place and familiar space may be confounded when the body is 'out of place', though spatio-temporal patterns may be quickly re-installed to reconfigure presence in a changed or unfamiliar space in order to regain ontological security.

Rhythms of mobility

A focus on the rhythms of mobility is sustained by the insistence that places are ceaselessly (re)constituted by flows and never reified or bounded. There are three senses in which the rhythms of mobility constitute place. Firstly, place is characterised by the mobilities that course through it, as in Lefebvre's (2004) account of the view from his Paris window where he espies the stop-start rhythms of pedestrians and traffic, and their variations over the diurnal cycle. Patterns of mobile flow thus contribute to the spatio-temporal character of place, whether dynamic or placid, fast or slow, and this is best ascertained at a still point from which mobile flows of varying tempo, pace and regularity are apparent. There is a regulatory dimension through which the braiding of multiple mobile rhythms is organised, with traffic lights and other apparatus, speed limits, highway codes, laws, road layout, and the dissemination of good habits reproduce familiar disciplinary conventions. As with other habitual, everyday enactions, such rhythmic systems are rarely apparent except when they break down or are violated, or where they no longer pertain. For instance, I have written about the variegated rhythms of many Indian roads, where a host of vehicles – lorries, buses, cars, bullock carts, bicycles, rickshaws – along with animals and pedestrians, all moving at greatly different speeds and styles, compose a far less regular rhythmic pattern than that familiar to British road-users (Edensor, 2000).

Secondly, regular rhythms of mobility, such as commuting (Edensor, 2009), produces a sense of mobile place. The speed, pace and periodicity of a habitual journey produces a stretched out, linear apprehension of place shaped by the form of a railway or road, and the qualities of the vehicle. Through commuting, a distinct embodied, material and sociable 'dwelling-in-motion' emerges (Sheller and Urry, 2006) as place is experienced as the predictable passing of familiar fixtures under the same and different conditions of travel, as with the commuters on Santiago's Metro system depicted by Paola Jiron (this volume), who gain pleasure from daily passage through familiar tube stations and scenic views. This stretched out, mobile belonging diverges from accounts that suggest that 'places marked by an abundance of mobility become *placeless*' realms of detachment (Cresswell, 2006: 31), for such assertions overlook 'the complex habitations, practices of dwelling, embodied relations, material presences, placings and hybrid subjectivities associated with movement through such spaces' (Merriman, 2004: 154). On the contrary, serial features install a sense of spatial belonging, including the road signage and roadside furniture that occur with rhythmic regularity, and the daily apprehension of routine features may provide a comforting reliability and mobile homeliness as consistent yet changing elements in a landscape. Journeys have a particular rhythmic shape. Particular phases reliably mark progress as traffic speeds up or slows and routinised elements such as the purchase of the daily newspaper or travel tickets enfold social relations into the daily ritual. Such sequential, rhythmic apprehensions also serve to highlight that which stands out from the norm: the sudden road accident, newly painted house or unusual bird; all stand out in sharp relief to the usual happenings. Paradoxically then, mobile experiences of place and belonging may be transient and fleeting as well as associated 'with prolonged or repeated movements, fixities, relations and dwellings' (Merriman, 2004: 146).

Thirdly, and enmeshed with the rhythms discussed above, the interior of a mobile vehicle or other form of transport is a different sort of place with its own rhythms. This is evident in the machinic pulses laid down by the engine and metronomic swish of windscreen wipers and indicators. Certain familiar and comfortable mobile environments lull drivers and passengers into a state of kinaesthetic and tactile relaxation – consider the beat of the train and the ticking over of the car engine at speed which together with a cushioned interior furniture facilitate a disposition to dream, listen to music or converse. Jiron also shows how daily travellers on the Santiago Metro impose their own rhythms of sociability, reverie, relaxation and independence inside the carriages, within a familiar spatial context enclosing fellow travellers, fixtures and signs.

A mobile sense of place is shaped by the mode and style of travel. For instance, Wunderlich compares 'purposive walking' at constant rhythmical and rapid pace' with the more varied rhythm of spontaneous 'discursive walking', as well as with the 'conceptual', critical walking mobilised by situationists and psychogeographers (2007: 37-8). These approaches to walking can, moreover, be combined experientially in a single walk, highlighting how one facet of mobile practices is the flow of continuous attachment and detachment to place, as exemplified by Labelle, who argues that 'walking may be a site for a radical placement and displacement of self, fixing and unfixing self to urban structures, locational politics and cultural form, locking down as well as opening up to the full view of potential horizons' (2008: 198).

Non-human rhythms

Recent ideas about the role of non-humans, ranging from objects, energies, flora and fauna within an expanded understanding of the social require us to account for the complex array of non-human rhythms that impose upon, exist separately and are entangled with human rhythms. Too often accounts have considered the non-human dimensions of place to be a passive backdrop upon which human activity unfolds. However, places are always becoming, and a human, whether stationary or travelling, is one element in a seething space pulsing with intersecting trajectories and temporalities. As Lefebvre says, '(There is) nothing inert in the *world*', which he illustrates with the examples of the seemingly quiescent garden that is suffused with the polyrhythms of 'trees, flowers, birds and insects' (2004: 17) and the forest, which 'moves in innumerable ways: the combined movements of the soil, the earth, the sun. Or the movements of the molecules and atoms that compose it' (20). Certain natural processes are 'only slow in relation to our time, to our body, the measure of rhythms' (ibid.). By acknowledging the usually cyclical rhythms of nature: processes of growth and decay, the surgings of rivers, the changes in the weather and the activities of animals and birds which breed, nest and migrate, we can identify the ubiquitous presences of non-human entities and energies in and through place. And as Rhys Evans and Alex Franklin show in this volume, combinations of living things – in their example, the horse and human rider – combine to produce rhythms that neither could achieve independently. The cycles of the moon and sun, and the millennial changes associated with climatic, geological and geomorphological events, possess rhythmic patterns and irregularities that deeply impact on place and space. Owain Jones highlights how the shifting rhythms of tide deeply mark particular places and are entangled in complex human practice as illustrated by literary and artistic works. At a longer temporal scale, such transformative phases rarely map onto temporalities of the human lifespan, though the rapid acceleration of climate change may require a shift in thinking rhythmically; indeed, as James Evans suggests, there is an urgent imperative to develop new economic and social rhythms which are better attuned to the quickly mutating ecological rhythms that signal impending catastrophe.

Everyday rhythms

Zeruvabel asserts that (1985: 2) 'through imposing a rhythmic "beat" on a vast array of major activities (including work, consumption and socialising), the week promotes the structuredness and orderliness of human life'. This also applies to the temporality of the day, for everyday life is constituted out of a multitude of habits, schedules and routines that lend to it an ontological predictability and security. Once learned and followed, these habitual procedures become unreflexive, are part of the way things are, though if the rhythm of the day should be disrupted and routines are thwarted, discomfort often ensues. The rhythmic structuring of the day is not merely individual but collective, and relies upon the synchronisation of practices that become part of how 'we' get things done.

Accordingly, the everyday greatly contributes to what Raymond Williams called a 'structure of feeling', a sense that emerges out of 'the most delicate and least tangible parts of our activity' (1961: 63), that generates a communal way of seeing the world in consistent terms, sharing a host of reference points which provide the basis for shared discursive and practical habits; as Stuart Hall puts it, 'forms of common sense which have taken root in and helped to shape popular life' (1996: 439). Clearly then, precisely because it is often beyond reflection and critique, the rhythmic lineaments of everyday life are weighted with power. According to (Carlson, 1996: 16), habits are 'discrete concretisation of cultural assumptions'. Frykman and Löfgren concur, stating that 'cultural community is often established by people together tackling the world around them with familiar manoeuvres' (1996: 10-11). For friends and family tend to share habitual routines and these 'familiar building blocks of body, family and kinship' are the basis for a wider sense of belonging (Herzfeld, 1997: 5-6), part of shared forms of 'habitus' (Bourdieu, 1986). I have written elsewhere of how these rhythms and habits may produce a sense of national identity, through the repetition of innumerable quotidian routines and habits, a cyclical ordering which organises, apportions, schedules and coordinates activities, producing an intuitive sense of synchronicity, which renders us aware of the simultaneous enactions of millions of others at a national level (Edensor, 2006). As Spillman notes, such rhythm 'flavours everyday life in familiar ways and … makes an unnoticed backdrop to public life' (1997: 2). In terms of scale, this can be applied to local and institutional contexts and increasingly, draw in global synchronicities.

Familiar places are the unquestioned settings for daily tasks, pleasures and rhythmically apprehended routines, with regular patterns of walking, driving, shopping and other routinised practices as part of familiar spatio-temporal experience. These patterns are marked by regular paths and points of spatial and temporal intersection which routinise action in space and collectively constitute the time-geographies (Gren, 2001) within which people's trajectories separate and cross in regular ways. Shops, bars, cafes, garages and so forth are meeting points at which individual paths congregate, providing geographies of communality and continuity within which social activities are co-coordinated and synchronised. This ongoing mapping of space through repetitive, collective choreographies of congregation, interaction, rest and relaxation produce situated rhythms through which time and space are stitched together to produce what Seamon (1980) calls 'place ballets' (also see Wunderlich, this volume). And the accumulation of repetitive events also becomes sedimented as individuals, through familiar bodily routines in local space, for instance, walk on tarmac pavements, patches of grass and wood and absorb these surfaces, forging 'a primary rivet aligning body with place' (Labelle, 2008: 189). By looking at the ways in which individuals find ways of latching onto environments through attuning themselves to musical rhythms, Labelle further depicts how people move through space in time with auditory rhythms, or whistle, sing or tap along; the rhythm simultaneously inside and outside their bodies. This, he argues, produces a personalised time that *supplements* placial and bodily rhythms and this latching on to particular beats, whether in the car or on a personal stereo (see Bull, 2000), or simply in one's head, aligns the body 'with a self-defined choreography' that generates 'links, stoppages, bolts and rivets to the existing architecture of time and space' (2008: 190). This can be extended to consider other gestures, daydreams and performances through which personal rhythms ground individuals in place.

The scales of rhythmic attunement to space vary enormously. For instance, at a national scale, space is arranged 'in a way that supports the bodily habits and routines of those who dwell there' (Young, 1997: 136), through the production of serial institutional fixtures, recurring signs and styles, familiar retail outlets and so on (Edensor, 2006). Yet where formerly reliable fixtures disappear, or movement to place beyond the national may produce alienation or a loss of belonging, such rhythms are suddenly absent. Yet as Mary Chamberlain avers in discussing Caribbean migrants to Britain, people are apt to restore familiar spaces, routines and timings. These migrants were prosaically

> concerned with the daily round of work, home and family. Perhaps it is precisely in the mundane that the culture of migration can be observed for it is within the everyday, within the family and the workplace, the home or the street, that family values and cultural practices are transmitted, contested, transformed and where identities evolve. (1999: 260)

A further interesting example of this quotidian recreation of familiar timespace is provided by O'Reilly's (2000) study of a British expatriate community in Spain who consolidate a shared identity in a space of otherness through enacting repetitive social rituals of Britishness.

It is also vital to acknowledge the role of technologies in affording rhythmic consistency. Schwanen points to the enrolment of non-human entities into rhythmic stabilities through the deployment of artefacts such as measurement devices, texts and automatic door closers. He argues that by delegating 'interaction and other relations into nonhuman entities, humans can give obduracy and transportability to those relations... they may routinize sequences of (human) acts' (2007: 12) along with other 'co-ordination tools' such as diaries, alarm clocks, planners that can manage the much-mooted 'time squeeze' (Southerton, 2003) that exemplifies the difficulty of managing work life balances (though see Potts, this volume).

The rhythmic rituals and habits discussed above produce places, landscapes and nations as recognisable and shared entities through everyday *synchronisation*, the simultaneous participation of millions of people in timetabled routines, which, according to Zerubavel (1985) is a fundamental principle of social organisation usually eluding analysis because of its very familiarity. The extent to which social life is marked by synchronisation is captured by Jarvis who remarks that 'most of us spend much of each day orchestrating continual movement in relation to others ... knowledge of where, when and how activities and relations are to be conducted is essential' (2005: 137). Such synchronicities are particularly exemplified and extended through mediated synchronicities which, though they have been represented as disembedding processes through which everyday face-to-face interaction is replaced by virtual simultaneous interaction, are typically re-embedded in the home (Moores, 1995). For the media has become part of the flow and rhythm of the everyday, as people switch their computers, mobile phones and televisions on and off. Benedict Anderson (1983) and Michael Billig (1995) have argued that the media fashion a sense of shared identity through which the nation can be imagined as that entity to which we belong, and the regular schedules and timetables of television seem to further ground this shared rhythmic alignment. For instance, Chris Barker cites a list of televised sporting events, political and royal ceremonies and soap operas that, he argues, 'all address me in my living room as part of a nation and situate me in the rhythms of a national calendar' (1999: 5-6). Moreover, other kinds of television programme, with their apparently casual and intimate presentation styles provide an 'illusion of presence or a feeling of familiarity' and a sense of 'simultaneity with the everyday lives of consumers' (Moores, 1995: 334). Roger Silverstone highlight how these television schedules organise household routines through which time 'is felt, lived and secured', producing repetitive viewing experiences 'embedded in the times of biography and the life-cycle, and in the times of institutions and societies themselves' (1994: 20). Although it may be argued that viewing habits, routines and rhythms are becoming increasingly customised with the expanding availability of numerous channels, in Britain at least, the largest audiences continue to be attracted by terrestrially broadcast soaps, sporting events, news programmes and dramas. Nevertheless, mediated rhythms clearly extend rhythmic synchronisation to a global scale. Maurice Roche draws attention to mega-events and 'the continuity of their role as national and international timekeepers and history markers' (2003: 119) and sporting events and breaking news stories are simultaneous witnessed worldwide (Brose, 2004).

Despite – or perhaps because of – this prevalence of quotidian ritual and routine, as Mels observes, rhythm 'disappears into the mist-enveloped reified realm of fixed things' (2004: 23). This book attempts to reveal these often unacknowledged pulses through which social life is regulated and experienced, yet as I discuss below, this does not mean that everyday life should be reified as solely ordered and predictable. For the everyday is a site for the enfolding of multiple rhythms, and though the immanence of experience is usually anchored by habit and routine apprehension, there is always a tension between the dynamic and vital, and the regular and reiterative.

Inscribing rhythm

I now explore how power is instantiated in attempts to regularise behaviours in accordance with particular prescribed rhythms. Lefebvre declares that for change to occur, 'a social group, a class or a caste must intervene by imprinting a rhythm on an era, be it through force or in an insinuating manner' (2004: 14). In the broader allocation and organisation of social time, 'choice is contingent upon material context, institutional regime and moral climate' (Jarvis, 2005: 150). Thus diffuse forms of power often seek rhythmic conformity and spatio-temporal consistency through the maintenance of normative rules and conventions about when particular practices should take place at particular times, for power 'knows how to utilise and manipulate time, dates, time-tables' (Lefebvre, 2004: 68). The rigour of school and work hours, regulations about commercial opening hours and the sale of alcohol, the time when 'noise' is labelled 'antisocial', along with many other beats that we are expected to follow, establish 'good habits', often laid down by the state (Frykman and Löfgren, 1996). The normative values associated with following 'proper' timings and rhythms is exemplified by McCrossen (2005), who shows how Sunday in the USA provided a moral marker in weekly time, encapsulating strictures about when to rest, work and play, and involving imperatives to follow slower rhythms on the Sabbath. The temporal parameters established by institutions and workplaces are especially tightly managed for those who must succumb to the schedule of the prison and asylum centre (see Conlon, this volume), where scope for laying one's own beat on official rhythms is more severely circumscribed.

The pathologising of 'erratic habits' in contrast to 'good habits', is captured by Highmore in his account of modernist regulators in Victorian London who identified deviant rhythms as 'wayward and unpredictable' (2002: 179). Clear contemporary parallels are evident in the supposedly wasteful rhythms of the long term unemployed and 'workshy'. There are, it is asserted, 'productive' and 'unproductive' rhythms (see Hall, this volume), the former privileging certain kinds of regular commercial endeavours and 'hard work'. Besides these weekly and daily scales, Elchardus and Smits also point to 'the timing and the duration of transitions and stages, or the chronological age at which (lifecourse) transitions take' (2006: 305), which they argue, follow idealised notions 'characterized by an unambiguous sequential order and a surprisingly strict timing of the transitions' (321). Alheit concurs, maintaining that life courses follow 'institutionalized expectation structures' (1994: 311).

In discussing the saturation of space and the everyday with dominant rhythms, many of these have emerged and are emerging from an ever-shifting capitalism. In terms of scale, rhythms vary from the forty- to sixty-year Kondratiev waves to shorter periods of economic boom and bust. John Tomlinson (2007) contrasts the slower, more fixed rhythms of earlier capitalist eras with the adaptable and flexible rhythms of 'liquid modernity'. Here, the rhythms of interest repayments, short-term and long term investments, cycles of innovation and obsolescence, rent payment, share fluctuations, product cycles and fashions are accelerating. The global sourcing of material and labour, just-in-time delivery systems and rapid networked informational and communication techniques characterise flexible production. In addition, the trading in futures and derivatives which has contributed to recent global economic turbulence has been facilitated by the speed of computer transactions and the gathering of instantaneous information amongst short-termist speculators within an unfettered market. Further, the 'increasing incidence of mergers, take-overs and make-over, rebranding, restructuring' and the mantra of 'managing change' institutionalised as a normative modus operandus (85) have also destabilised working life and career structures. This seems to characterise an even more intensified process through which there is 'the perpetual disruption of temporal and spatial rhythms' (Harvey, 1989: 216) and this is very much manifested in the landscape and place with cycles of ruination, construction and regeneration. However, there are different temporal perspectives adopted by different actors and for different strategies and endeavours. Stock exchange activity typically focuses upon short term activity, banks may mix short and long term investment strategies, and corporations often adopt long-term planning (Brose, 2004) .

Modern industrial rhythms were typified, of course, by Taylorism, through which production-line processes were spatialised in the factory. Ideally, a repetitive, optimal sequence of events from shop floor to packaging department to loading bay was strictly timed according to the clock to instantiate highly ordered rhythms to which workers had to to conform, the logic of which results in ever shorter periods of time being carved up into utilisable segments. Practically, clock time provides a regular measure by which employers can establish rates of pay for workers over days, months and years. Despite the imperative for workers to commit themselves to this metronomic labour, entrepreneurs have little hesitation in abandoning a site of production if location elsewhere will maximise profit. In this sense then, normative working rhythms can be instantly destroyed by the rhythms of accumulation (Castree, 2009). In addition, there is always the incipient tendency for entropy, subversion, breakdown and thus, arrhythmia. Yet capital is ever-ready to instantiate new forms of temporal regulation. For instance, Helen Jarvis identifies the 'work-centredness of contemporary life associated with the new economy' (2005: 134), and Tracey Potts argues that for certain professionals, lifestyle is increasingly matched 'to the beat of workstyle' (2008: 100) so that domestic drills and leisure become complementary to work routines whereas previous conceptions understood leisure as a relaxed lapse into an unpurposive tempo which was the opposite of work. For these workers, Southerton contends that the change towards flexible and individual rhythms through a reflexive self-management and away from the 'collectively-maintained temporal rhythms' of yesteryear (2003: 7) paradoxically intensifies a sense of being hurried and short of time. Yet as Jarvis also insists, we must beware of hyperbole since 'the harriedness associated with daily life in cities like London today is not universal but unevenly manifest' (2005: 134).

The rhythms of space also change with the acceleration of the rhythms of consumption, fashion and innovation under an intensified consumer capitalism. The shortening of product life cycles, accelerated obsolescence in products, the rapidity with which fashions become outmoded and are replaced by trendier items, and other strategies that feed the desire for distinction mean that rhythms of marketing, advertising and acquisition have sped up for many. One process through which rhythms of consumption become inscribed in place is depicted by Mattias Kärrholm (2009), who shows how retailers in Malmo, Sweden, attempt to synchronise commercial rhythms with everyday urban rhythms and mobilities. He argues that as commercial activities and retail areas have become more predominant in urban space, public life increasingly take place within shops, pedestrian precincts and shopping malls, not to mention previously uncommercial spaces such as museums, libraries and airports. This commercial consolidation involves temporal as well as spatial control, and is producing new strategies for urban synchronicities through which flows and rhythms are framed in timespace to bringing people and activities together. Places in which this increasingly normative temporal and spatial territorialisation is absent may be regarded as deficient. Shops both instantiate urban rhythms through their opening times, which are becoming extended, and through their alignment with other rhythmic activities, such as attending cultural events and festivals, and taking holidays. The imperative to maximise commercial interests means that business may tend to turn polyrhythmic landscapes into isorhythmic ones, through which other rhythms are orchestrated to coincide with the rhythms of shopping. The alignment of the rhythms of commodification and their configuration with other everyday rhythms is also cited by Anne Cronin (2006) who shows how the urban landscape is increasingly saturated with advertisements precisely at points of dense mobile passage and temporary cessation of traffic flow.

Lefebvre emphasises the tendency for the rhythms of capitalism to pervade the everyday, commodifying previously untapped areas of quotidian experience, setting down the beats through which work and leisure proceed, and thus rendering the everyday banal and meaningless, an alienated realm. Yet despite this, as Highmore puts it, 'secreted within the everyday were the elemental demands for everyday life to become something other (something more) than bureaucratic and commodified culture allowed' (2002: 113) as I now discuss.

The illusion of rhythmic order

Despite the prevalence of self-managed routines and the imposition of bureaucratic, state and capitalist rhythms upon individuals in the realm of the everyday, it is important to avoid the inference that the quotidian is thus a sphere of entrapment and stasis. Instead, as Gardiner declares, the everyday can be 'polydimensional: fluid, ambivalent and labile' (2000: 6) for 'in the everyday enactment of the world there is always immanent potential for new possibilities of life' (Harrison, 2000:

498). Quotidian existence can be open-ended, fluid and generative, a sensual form of experiencing and understanding that is 'constantly attaching, weaving and disconnecting; constantly mutating and creating' (502). Lefebvre is explicit on this subject, recognising that 'to become insomniac, love-struck or bulimic is to enter into another everydayness' (2004: 75). Developing this theme, Simpson argues that rhythm provides a backdrop to life against which the usual and the unusual unfold, against which practice is reproduced and improvised, (re)producing 'a stable but also evental everyday temporality' replete with 'different durations with concomitant durabilities' (2008: 810). Rather than considering linear rhythms to be constituted by repetitive and oppressive routine, he considers them to form a field of potential. In investigating street performance in London's Covent Garden, he notes the linear rhythms of the institutional strictures placed on the structure of the shows, the embodied rhythms of performers and audience, seasonal and climatic rhythms, and the rhythms inherent in the performances, while also noting the arrhythmic events of a downpour and the sun's heat. He thus highlights both predictable and contingent elements that play out during one everyday event. While the character of the everyday does not necessarily engender a romantic resistance to power it is characterised by immanent and emergent possibilities as well as repetitive rhythms. Lefebvre acknowledges the linear monotony of 'the daily grind, the routine' (2004: 30), but argues that cyclical rhythms are wrapped 'in rhythms of mental and social function' (8), a melding which ensures that the linear can never be wholly triumphant, for in a polyrhythmic assemblage, rhythms influence each other, sometimes achieving eurythmia, where stability persists, and sometimes, arrhythmia, where they jar and clash.

A useful metaphor for everyday experience is 'flow', which I conceive as a sequential process through which immanent experience is replete with successive moments of regular attunements to the familiar and the surprising and contingent, and is typified by periods of self-consciousness followed by subsumption into unreflexive states once more. Lefebvre conceives moments of surprise, insight, revelation and sharp self awareness, as occurrences which cannot be appropriated through commodification, and which thus have the potential to serve as opportunities to escape everyday alienation, though they are difficult to express and bring to conscious thought (see Seigworth and Gardiner, 2004)

A further element here is that the linear structuring of social time can make it productive. Rhythmic routines provide a framework through which to economise on life, and establish conditions for the unreflexive, playful, subversive and improvisational to emerge. Accordingly, Crang's (2001) insistence on the always emergent nature of a present conditioned by the continuous influence of past and present need not obviate the presence of habitual rhythm, for as Lefebvre asserts, rhythm must be understood as 'an aspect of movement and a becoming' (1996: 230). He further claims that 'there is always something new and unforeseen that introduces itself into the repetitiveness: difference' (2004: 6). This is akin to Deleuze and Guattari's (1987) notion of the *refrain*, a repetition that produces familiarity, a territorialisation that militates against chaos. The sociality of the refrain inheres in its transference to others, who articulate the repetition while also slightly changing it. This always dynamic process provides a recognisable, familiar, inter-relational sense of place and practice that echoes through the actions of many but also allows for improvisation and adaptation. Crang here misrecognises Lefebvre's rhythmanalysis as constraining, oppressive and repetitive in contradistinction to an ever-new becoming in which past, present and future always occur and combine in unique ways to make new temporalities. This insistence on the potential of routine is also echoed by Hallam and Ingold who argue that in repetitious practices there 'is creativity even and especially in the maintenance of an established tradition ... (for) traditions have to be worked to be sustained' (2007: 5-6).

The productiveness of repetition is exemplified by the gaining of expertise so that practice becomes automatic and easy, or allows moments of eurythmy to emerge. This is depicted in Dydia Delyser's (this volume) discussion of the ways in which training – or dressage – might prepare aeroplane pilots for emergencies while airborne, but in the event of such arrhythmias of engine and body and the consequent peril, a heightened attunement to circumstances and ability to improvise are vital, and they emerge from training but are not determined by it. Similarly, Evans and Franklin (this volume) show how the synchronising of rhythms between horse and rider can produce eurhythmic moments of 'floating harmony' during the equestrian pursuit of Dressage, as trained bodily rhythms merge into transcendent moments of sensation and accomplishment. Here the endless, repetitive, everyday dressage of rhythmic action bears fruit in a shared euphoria between horse and rider.

It is important therefore, to avoid assumptions that managed normative rhythms possess an overarching force that compels individuals to march to their beat. Instead, people are apt to attune themselves to the rhythmicity of the moment through breathing, gestures, pace of movement and speech. And as Labelle declares:

> To locate one's own time is to derive a personal spacing within the built; it is to cut into the standardisations of daily routine an interval by which to fashion perspective, according to the mutations and nuances of time. (2008: 193)

Besides this personal engagement with the rhythms of place, putting one's own beat in space, the effort required to maintain rhythmic and temporal order should not be underestimated. Consistency is always emergent and contingent, reproduced again and again. It is why the endless maintenance of the material and social world depends upon the rhythms of maintenance to produce continuity and stability (Graham and Thrift, 2007), the army of repetitive cleaners, technicians, managers and police who ensure that matter, activities, humans and non-humans remain in their 'proper' place. Several chapters in this book highlight the impossibility of maintaining strict rhythmic and spatial order. Richard Hornsey highlights how with the advent of the motor car, Taylorism influenced attempts to impose new rhythms on the pedestrians of 1920s and 1930s London to prevent disruption to the flow of vehicular traffic and minimise accidents, and yet such rigorous training proved to be partial, for pedestrians are always likely to disregard rules, to improvise, to make their own ways across the road. Similarly, Monica Degen (this volume and 2008) shows how regulatory processes of gentrification are only partial in their imposition of homogeneous rhythms in El Raval, Barcelona, and in any case, the serial nature of many gentrified spaces causes inhabitants and tourists alike to be attracted to sites where rhythms are more variegated, with an array of cultural practices, sensations and spaces complementing the adjacent, more homogeneous spaces of consumption. Though regeneration professionals are opportunistically utilising this rhythmic diversity to market place in a more sophisticated manner, the cultural adaptation and resistance that produces a rhythmic dynamism to place stays ahead of official designs.

The fragility and transience of rhythmic routines is discussed by Alheit (1994) who identifies how disruptions to rhythmic regularity such as divorce or the birth of a child, may mark a new phase in life, or greater events, such as the advent of illness, war or unemployment, potentially causes instability and anomie without secure and consistent temporal tethering. More cheerfully, carnival can be seen as an exemplary time when everyday rhythms, like other ordering principles, are suspended. Other sensual experiences have the facility to similarly replace normative rhythms, as Jackson discusses with regard to music festivals, pointing to how '(T)he deeply sensual quality of the bass, its material succulence, penetrates the flesh like an alternative heartbeat that initiates a new physicality' (2004: 29). He further highlights how 'intoxication, the crowds and the music all work together to create new physical and emotional rhythms that underpin an alternative experience of self-in-world' (Jackson, 2004: 30).

Frykman and Löfgren argue that under conditions of globalisation, the disruption of familiar rhythms has become more commonplace

> in a mobile culture where people constantly meet otherness, habits are brought to the surface, becoming manifest and thereby challenged. It is precisely because people in their everyday lives meet different habits that they are forced to verbalise and make conscious the things that are otherwise taken for granted and thus invisible. Once a habit has been described, it has also become something on which one must take up a stance, whether to kick the habit or to stick tenaciously to it. (1996: 14)

This brings us to *resistant* rhythms, which proffer alternative modes of spending time, different pacings and pulses which critique normative, disciplinary rhythms and offer unconventional, sometimes utopian visions of different temporalities. Indeed, resistance to the rhythms of capital have always accompanied rhythmic imposition, and prefiguring the discussion about the speed-up of capitalist rhythms through global processes below, Highmore insists that emphasis should also be placed 'on slower, more sedentary and stagnant aspects of modern life' (2002: 172), citing the work-to-rule and the go slow, to which we may add ideas about 'chilling out' and hedonistic drug-taking as rhythms counter to modern speed (2002: 175). The disruption of accepted rhythms – a motionless body in a crowd, a blockage in the road, a power cut – makes apparent that those rhythms that usually hold sway have been thwarted – and noisy nightime-revellers, loiterers and mall rats, cyclists riding up a one way street (see Spinney, this volume) or convoys of fuel protesters more overtly resist the ordered rhythmic flows laid down by the powerful.

An increasing number of people also question and resist certain work and consumption rhythms that might be characterised as the rhythm of the 'rat race'. Neo-tribes, dole-ites and idlers are evident blockers of rhythmic flow. Shaw (1991) discusses how less adversarial rhythms are sought by those who seek out slowness during holidays or as part of 'downshifting', moving to places that they believe offer slower everyday rhythms, such as the countryside or non-western destinations. And those who have been pressed to enter a faster life and workstyle pulse may cling to older rhythms. For instance, Rau shows how following German unification, East German enterprises were unable to keep up with the increasing pace of western capitalist production after having encouraged 'a very special working time culture that promoted task-related creativity and cooperation rather than strict adherence to temporal rules' (2002: 281), and workers from the former GDR resisted these unfamiliar rhythms. Such examples resonate with the flaneur's habit of taking turtles for a walk to ensure they resisted the accelerating pace of 19th century Paris. Meadow's insomniacs (this volume), supposedly arrhythmic in their inability to harmonise with dominant diurnal rhythms, are reconceptualised as possessing the a heightened watchfulness or unconventional critical perspective.

The most obvious form of resistance to normative rhythms is the 'slow' movement, who in championing the values of authenticity, conviviality, variety and local distinctiveness declare in their manifesto, 'we are aiming at those who wish to listen to the rhythm of their own lives, and possibly adjust it' (Capatti, 1996: 5). This negotiation with everyday rhythms and 'a commitment to occupy time more attentively' (Parkins, 2004: 364) aims to gain enhanced aesthetic or sensory experience and ensure that spaces of slowness supplement fast spaces in public and private domains. This is spatially manifest in the usually small cities that have signed up to the Cittàslow movement which involves engaging more slowly 'in everyday practice in particular ways at a routine, personal, individual level' (Pink, 2007: 63).

Hyperbole and fast rhythms

Finally, having discussed the slow movement as a form of resistance against the supposed speed-up of everyday life, without minimising the impact of certain accelerating rhythms, I want to question whether rhythms are speeding up as much as is often claimed. Tomlinson (2007) argues that a culture of immediacy has emerged as the deployment of technology has aimed to reduce effort, journey and contact time, and has introduced expectations of instant gratification, notably through consumption and the rapidity with which fashions succeed each other, and has achieved this through access to credit systems, with relationships, musical tastes and commodities all increasingly dispensable. Tomlinson also portrays the fast life as predicated upon intensity of experience, a disposition to cram experiences into life, epitomised by the heroic phrase, 'live fast, die young'. Dodgson further explains that the speed of disembedded flows of information is frequently held to result in a 'permanent breakdown in the flow of meaning from past to present, a break that would have us living in a perpetually disjunctured, isolated present' (2008: 10). Objections to this hyperbole may start with the obvious point that speed can lead to slowness, that the more haste, the less speed in certain circumstances, for instance in the traffic jams that result from too many car-drivers wanting to move too quickly at once. More important however, is the dismissal of the continual presence of routine and habit, despite the velocity, fast rhythm and pace of flows. As is clear from the discussion of the everyday above, despite claims that in postmodern times all is fluid, as Jenkins asserts, 'the world of humans is not experienced by most people as chaos, perpetual ferment and novelty, evanescence and fleeting immediacy' (2002: 269). By focusing on the pulse and tempo of stretched out flows, such commentators ignore the denser, differently scaled networks through which time-space is experienced and organised, often in what Ellegård and Vilhelmson call a permeable 'pocket of local order' which acknowledges the 'significance of stationarity, stability and proximity in the use of place' (2004: 282). Over-general accounts of a frictionless, speeded-up world in which all is mobile fails to account for the unevenness through which access to speedy communication and travel is acquired. Moreover, there is a hint of the ahistorical in the assumption that all is new and utterly changed. As Conlon points out in this volume, the everyday also carries with it residual rhythms from long ago.

This is not, however, to suggest that it is unimportant to account for certain processes that produce a speeding-up of experience and practice. In fact, rhythmanalysis is an exemplary method through which to mediate between overdrawn, static reifications of place hyperbolic accounts about spaces of flows. Tom Mels cites the contribution rhythmanalysis can make to a 'reanimation of people, things and places' (2004: 10), avoiding reification and acknowledging the dynamic temporality of the world and the processual character of place. Rhythmanalysis welcomes vitalist perspectives but warns that not everything continuously changes. It recognises consistencies, repetitions and reproductions, moments of quietitude, not withstanding the furious work that goes into the sustenance of stable arrangements, and is open to moments of chaos, dissonance and breakdown; moments of arrhythmia.

PART I
Power and the Rhythms of Place

Chapter 2
Consuming Urban Rhythms: Let's Ravalejar

Monica Degen

During 2006, while walking through the neighbourhood of El Raval, a poster in a small independent shop captured my attention as it proclaimed in Catalan, Spanish, Urdu and Tagalog (El Raval's most spoken languages): *ravalejar* – in other words 'to do, to live El Raval'. It was the first time I had seen a place turned into a verb, into an activity which inferred that by being in or walking across El Raval one could partake in the neighbourhood's life, merge with it, and 'do' El Raval. Of course one could easily dismiss this as just a clever marketing campaign promoted by Barcelona's city council and local shop-owners. However, as I suggest in this chapter, there is more at stake here, and presenting El Raval's daily life as a set of experiences to digest or immerse oneself within, can be regarded as part of a growing trend to attempt to control urban experience and commodify urban rhythms.

El Raval is a neighbourhood that has become paradigmatic for entrepreneurial urban regeneration processes in cities across the world. Since the 1980s, Barcelona's city council has invested large amounts of money in redesigning this working class, and former red light district, into a cultural quarter to dispel its negative reputation, the ultimate aim being to include valuable city centre real estate into Barcelona's middle class and tourism circuit. The general upgrading of El Raval's housing stock and public spaces was accompanied by a range of flagship developments over the years. Starting with a 'starkitect' modern art museum designed by Richard Meier, built in 1995 to spearhead a new 'cultural quarter' in the north of the neighbourhood, the re-organisation of El Raval's spatial landscape continued in 2000 with the construction of a new avenue: La Rambla del Raval, that in Hausmannian fashion cut through the heart of the neighbourhood's main prostitution and drug trade area and required the demolition of more than three blocks of low rent apartments. The cultural re-signification of El Raval has been further supported institutionally by attracting a range of university faculties, research centres and cultural institutions into the once dilapidated district, sometimes housing them in historical buildings, at other times building them anew. Since the late 1990s a string of art galleries, restaurants and designer boutiques have moved into El Raval gradually replacing old neighbourhood cafes, brothels and the neighbourhood's manual industry. So far, this could be regarded as a familiar tale of gentrification processes, only that in El Raval one of the more unexpected outcomes in the late 1990s has been the constant influx and settling of non-European migration, establishing El Raval as Barcelona's most multicultural area. As a result the gentrification of El Raval is far from complete and, instead, the neighbourhood offers an eclectic mixture of

Figure 2.1   Ravelejar

spaces where minimalist designer boutiques live next to halal butcher shops and
Filipino hairdressers, and where the last vestiges of cavernous neighbourhood bars
with old gentlemen playing dominoes and drinking 'carajillos' stand their ground
against luminous pink lit cocktail bars.

The remodelling of a marginal area into a cultural quarter involves a dramatic
transformation of the urban spatial structure and leads to a deep change of its
experiential landscape as its urban fabric and social uses are altered (Degen,
2008). Let me explain this in more detail. Regeneration strategies entail the

dismembering and re-assembling of the built environment. In such processes, buildings get whitewashed or demolished, streets repaved and widened, new shops and attractions etched onto a re-signified urban landscape. The effect is the formation of a novel social geography of place as new social groups enter the area, sometimes replacing old inhabitants, yet at other times, living side by side with the old residents. An inevitable consequence is that the activity and sensory rhythms of a place change and are reorganised. Essentially, urban regeneration processes transform the sensory qualities of places which in turn shape the exclusion or inclusion of certain cultural practices and expressions in the public life of the city. A feature often ignored in the literature that assesses contemporary urban renewal processes is that these changes occur progressively, over time, for the process of regeneration is gradual, with the consequence that a locality adapts, refractures, reworks and at times even discards these regeneration processes as the example of El Raval illustrates. So, how are we to research these elusive experiential expressions and negotiations of spatial power relations? I argue that an analysis of sensory rhythms in urban public places reveals the various and contested ways in which place experience is created, controlled, consumed, or commodified. I begin with a discussion of the relationship between sensory embodied experience, urban rhythms and urban change. In the second half of the chapter I examine the transformation of El Raval's experiential make up since its urban renewal in the 1980s. Firstly, I show how in the first instance the diverse regeneration processes were an attempt to control and sometimes erase what were considered negative and unruly urban rhythms. Secondly, I argue that as these attempts have failed, an emergent and distinct experiential geography has become a central ingredient in El Raval's place marketing. My discussion draws on continuous ethnographic fieldwork conducted in El Raval since 1998.

Senses, rhythms and urban change

The senses mediate our contact with the world. In urban environments, this means that public life is first and foremost experienced through the sensory body, a feature often forgotten in a dominantly occularcentric Western society (Degen, Rose and Basdas, 2009). We do not only see but feel and hear the cold aura of an empty granite square. And, not only can we see the buzz of a busy pedestrian street, but we sense bodies brushing past us, hear and smell bodies, perfumes, activity. To put it simply, cities and bodies are mutually constitutive: '[...] the form, structure, and norms of the city seep into and effect all the other elements that go into the constitution of corporeality and/as subjectivity' (Grosz, 1998: 47). The built environment, the shops and social life we encounter, people's everyday practices, all amalgamate through our embodied perception to create a sense of place, or what Lefebvre has described as the 'lived space': the concrete, subjective space of users, the space of everyday activities where 'the private realm asserts itself, albeit more or less vigorously, and always in a conflictual way, against the public one' (1991: 362). The movement of bodies through spaces generates an ephemeral, continually changing and fluid space experience. For Allen therefore a city is filled with 'expressive meanings [that] have more to do with how the city is felt than how it is perceived or conceived' (1999: 81). The diverse combinations of material and social features produce 'felt intensities' (Allen, 1999) and come together in what is experienced as the ever fluctuating nature of public life which, I argue, can be understood through Lefebvre's (1991) concept of rhythmanalysis. While some writing has emphasised the temporal aspects of rhythms (Crang, 2001; Elden, 2004), less attention has been given to Lefebvre's quest 'to rehabilitate sensory perception' (Meyer, 2008: 149) in our understanding of the urban.

Towards the end of his life Lefebvre became increasingly interested in the senses, especially in their role in establishing spatio-temporal relationships between the body and space. His premise was that social space is experienced first and foremost through the body. Rhythmanalysis plays an intrinsic part in exposing the social production of space for Lefebvre. Indeed rhythmanalysis seeks to capture empirically the embeddedness of social relations in the sensory make up of space. Hence the rhythmanalyst, 'must simultaneously catch a rhythm and perceive it within the whole, in the same way as non-analysts, people, *perceive* it. He must arrive at the *concrete* through experience' (Lefebvre, 2004: 21). The rhythmanalyst does not restrict his/her observations to the visual but listens out, experiences movements in everyday life, the cyclical comings and goings of people, of nature, the subtle transformations of space. One observes and perceives to attain a particular state of awareness (Allen 1999), to make sense of the ever changing character of place. The analysis of rhythms attempts to capture the temporal and lived character of space. Similar to an orchestra that builds a symphony from different instruments, each one playing its own tune to its own rhythm, we can imagine an urban environment in which the interplay of multi-layered perceptions (the tune) and different intensities of these (the rhythm), create a sense of place. For Lefebvre, there are many different rhythms in the city, from the flows and stops of car and pedestrian traffic to the more subtle rhythms of the changing seasons or the body's individual cycles. To understand how urban change produces new sensuous geographies there are two particular rhythms one needs to pay attention to; firstly, rhythms in terms of activity/movement and secondly, sensuous rhythms that relate to our embodied, sensory experiences of place.

Activity rhythms are created by the daily movements, everyday, repetitive spatial practices: the coming and goings of people to and from work, the rush hour traffic, lunchtime breaks, the garbage men collecting rubbish, that Jane Jacobs (1961) famously describes as a 'sidewalk ballet'. The mix of superimposed, parallel flows that repeat each day confer place with a specific rhythm and give those who live, visit and work there a sense of location. It is this repetitiveness that produces distinct spatial rhythms.

A focus on sensory rhythms on the other hand, helps us to understand how spatial experience is being shaped by constantly shifting sensescapes that easily slip through our fingers and we tend to summarise as the 'feel' or 'atmosphere' of places. Activity rhythms are intricately linked to sensory rhythms. Public life is punctuated and produced through activities we experience through the senses; thus we hear the buzzy droning of mopeds, feel the pressure of other bodies in the lunchtime crowd, notice the stale smell of recycled air in the underground or see an army of street-cleaners descending in the early morning hours onto the city's empty streets. However, other sensory rhythms surrounding us are afforded by the physicality of spaces. For example, the visual monumentality of a church, the touchscapes reflected in the various textures surrounding us: the coarse touch of brick, the cold feel of metal, the soft yet grimy seats of the nightbus, the wet scent of rainy streets or the shiny and hard marble in a shopping mall. To understand daily urban life and experience as constituted through a layering and multiplicity of rhythms helps to conceive the city as a polyrhythmic ensemble, 'the idea of the urban not as a single abstract temporality but as the site where multiple temporalities collide' (Crang, 2001: 189). Moreover, as Crang (2001) argues, the rhythmic city is based on an understanding of time-space relations as a continuous folding and unfolding of individual past and future experiences with the multiple temporalities inscribed in the surfaces of the city. The material expression of these temporalities becomes especially poignant in areas undergoing regeneration where the temporality of decay and regeneration is mapped onto the urban texture, producing both temporal and sensory juxtapositions.

Sensescapes fluctuate in intensity and in their relationships. What interests us especially about the fluctuations of rhythms is that once the rhythmanalyst has established the interaction of rhythms, the next step is to determine what kind of relationship these rhythms have and to 'keep [his] ear open', although he does not only hear words, speeches, noises and sounds for he is able to listen to a house, a street, a city as he listens to a symphony or an opera. Of course he seeks to find out how this music is composed, who plays it and for whom' (Lefebvre, 1996: 229). We can see here how Lefebvre clearly identifies power relations as a crucial part of the production of urban rhythms. In fact, sensory perceptions are far from neutral but, as several critics have highlighted 'social ideologies [are] conveyed through sensory values and practices' (Howes, 2005: 4; see also Law, 2001, 2005; Edensor, 2005). To understand the cultural geographies, meanings, values and practices of places it is paramount to interrogate whose rhythms intensify, alter or disappear as different social groups make their claim to space.

Domesticating El Raval

Spatially, El Raval (literally meaning periphery or outskirt) has always been, both physically and symbolically, at the margins of and in opposition to the bourgeois city of Barcelona. Built between two city walls, it became the cradle of Barcelona's textile industry in the 17th century and Europe's densest working class neighbourhood. When the industry moved out at the end of the 19th century the empty factory shells were transformed by waves of internal Spanish migrants into precarious living areas which led to a flourishing of non-licensed residences and subletting as well as the construction of shacks on most rooftops, the so called 'barraquismo vertical' (vertical slums). At the start of the 20th century, its geographical proximity to the harbour saw it transformed into Barcelona's main entertainment and red light district. Cafes, taverns and music halls stood side by side with brothels and sex-shops providing it with a unique bohemian and cosmopolitan character which attracted artists and punters alike. Soon the area became infamously known as 'Barrio Chino', because its street-life resembled the seedy Chinatowns of North America.

After the Spanish Civil War (1936-1939) the neighbourhood increasingly deteriorated as a new moral climate led to the closing of theatres and other leisure establishments (in 1956 prostitution was officially banned by the Spanish government). During the 1960s new social trends moved the nightlife gradually to other parts of the city. Gradually El Raval decayed into an area of cheap prostitution and sordid sex shops. The neighbourhood became increasingly marginalised in the social imaginary. The final straw was the entrance of heroin into the neighbourhood in the 1980s which led to an extreme level of insecurity and a further deterioration of El Raval's social life that forced many residents to flee the neighbourhood.

It was at this time that the first plans for urban renewal were devised with the main aims to improve the quality of life of its residents and control and domesticate what was perceived as an unruly public life and decadent sensory landscape. It was a place that most of Barcelona's population would shun, segregated spatially from the rest of the city by wide avenues and hidden behind tall 'panel buildings' – increasing its isolation from the city centre. From an urban planning perspective this was an area that had been largely untouched by the organising power of modernity. While El Raval's poor housing conditions had inspired the urban planner Ildefons Cerdà to develop the rational grid of the Eixample in 1854 (Barcelona's modern urban expansion which led to the demolition of the old city walls), El Raval's squalid landscape had received no major investment or restructuring during the 20th century. As a result, at the start of its regeneration in the 1980s El Raval's urban landscape was still based on a chaotic medieval street pattern, with narrow streets, a multitude of courtyards and alleys bordered by grey five storey buildings, closed shop shutters, boarded up windows and crumbling walls. On the streets the musty smell of abandonment mixed with stinging waves of urine. Occasionally, a glimpse into a workshop revealed a carpenter engaged in carefully polishing a chair and provided a glimpse of the lively working class neighbourhood El Raval had once been. On its streets, one would come across old women returning with trolleys from the market, a young man scavenging for food in a bin and a haggard prostitute leaning against an empty doorway.

The first step of the regeneration was thus to 'air' El Raval, to open up what had been regarded as a closed, insidious environment by cutting different sized airholes into the dense space, a method described as 'esponjamiento'. One could view this as a modernist planning project of ordering, reorganising and cleansing spatially and sensuously a pre-modern space. Hence, we witness the demolition of many houses to be replaced by a multitude of new public squares and large open spaces that allow the eye to roam, such as the area around the new Museum of Contemporary Art. The overall design style of these schemes is European modern. Wide and pale surfaces, the use of glass, steel, and granite provide a coherent textural mix which would be equally at home in Berlin or Stockholm. These spaces are characterised by a sensuous uniformity: noise gets filtered through the spaciousness, smells quickly whisk away, tactility is minimised by the smooth surfaces. A hierarchical relation of the senses is afforded by the spatial design of the environment in which the sensuous rhythms of the place heighten visual apprehension whereas odours, sounds and tactile experiences are relegated to supporting features. Cultural critics have identified a clear shift in contemporary cities towards producing a recognisable global iconography in which buildings by global architects have become status symbols of their post-industrial success and 'connote ideas such as cosmopolitanism, globalism and designer status' (Smith, 2005: 413; see also Evans, 2003). This sensuous and spatial standardisation makes recognizable environments for tourists, self-consciously designed spaces that produce familiar sensations and draw on common cultural capital. They fit into Zukin's (1995) description of commercialised spaces for visual consumption, to be captured and circulated in travel magazines and tourist guides.

During the re-assembling of El Raval in the late 1990s, the continuous humming of bulldozers, interspersed with the growling noise of demolition, became a regular tune. Leftover walls of gutted houses revealed the shadow of a bed-frame, the blue tiles of washbasins were all that remained of the communal toilets – physical ghosts of disappeared voices – and the roaring sound of construction was foretelling its own story of new apartments and broad avenues. A geographical segregation of sensory rhythms started to emerge as tourists, newcomers and museum workers accessed the cultural attractions through specific streets which were progressively transformed into light 'regeneration corridors' lined with cleansed heritage, modern cafes, libraries and boutique hotels. Other adjacent streets were characterised by grimy walls and closed shutters.

At first, the new spaces, especially the Plaça dels Angels surrounding the Museum of Contemporary Art, stood in stark contrast to the rest of the neighbourhood. Yet, very quickly these spaces were appropriated by the residents of El Raval. Dog owners walked around in the early morning, their dogs defecating on the granite floor, children started playing football and skating on its slopes, Filipino and Moroccan families gathered for picnics in the cool breeze of the evening and as soon the museum doors closed, homeless people assembled their card-board shelters in the corners. Hence, the purity of the visual became sullied by manifold sounds and smells as the spectacular space was digested within the neighbourhood's daily life. To counter some of these 'less desirable' activities and to regulate the urban rhythms, the council resorted to strategies such as the promotion of late opening hours for galleries, museums and bookshops and the organisation of a various festivals, markets and fashion shows in these new public places.

Consuming and commodifying El Raval

The start of a second phase of regeneration was marked by the opening of the new Rambla del Raval in 2000, initially a long empty avenue lined by stubby palm trees. This signalled a more mature phase of urban planning which started to emphasise preservation over demolition. More importantly, as mentioned earlier, El Raval's population has radically changed since 1995 as an increasing number of European and non-European migrants have settled and now comprise 48 per cent of the population. Slowly Catalan corner shops have been replaced by Pakistani delis that are open until late at night, the sound of Filipino dialects emanates from call-centres, the gradual emergence of mosques and prayer halls are signs of a parallel sensory re-ordering of sensuous and activity rhythms alongside the forces of regeneration.

El Raval's urban transformation is thus sensed in diverse ways and at different paces around the neighbourhood. Thus, on the same street one can observe the sand-blasting of 19th century buildings, listen to the roaring fall of rubble of yet another demolition, and hear the sound of hip-hop music coming out of a new record shop. A diversity of sensuous rhythms and intensities have started to blend, so that 'the aroma of chic [has begun] to waft through the once-pungent streets of the neighbourhood' (Richardson, 2004: 135). The mixture of ethnic cultural activities, 'cool' venues and traditional working class life has started to map an array of sensuous juxtapositions. Despite the planners attempt to control El Raval's public life, the neighbourhood is developing its own melody.

Informal interviews held with new younger residents and users of El Raval at the end of the 1990s revealed that it was precisely this mixture of new and old; the unregulated sensory combinations emerging from the various social groups living here – bohemian gentrifiers, migrants and leftover poor – that attracted them in the first instance to live in this neighbourhood. Both newcomers and tourists celebrate the 'multicultural' character of the area by favouring the sensory bricolage available in its public spaces to a temporally and ethnically homogenous Barcelona. Furthermore, a common impression was that El Raval's reputation as a marginal place had isolated it from the rest of the city, thereby preserving forms of sociality that have disappeared in the 'modern city'. For example it is still common in old, traditional shops to have chairs on which customers may sit, wait to be served and have a chat. These traditions are continued in many immigrant shops. Such forms of sociality lead El Raval to have its own aura and temporality, creating, as many residents describe it, 'a village within the city'.

What we see here is how El Raval and its inhabitants are producing their own individual melody out of a precarious and momentary balance of middle class taste, working class lives, gentrification processes and immigration. As Hannigan (2004) and Molotch et al. (2000) comment, place distinctiveness is hard to engineer; instead a distinctive urban tradition 'arises through interactive layering and active enrolments over time, something that is difficult to produce all at once' (Molotch, in Hannigan, 2004). At present, El Raval is a polyrhythmical neighbourhood 'composed of various rhythms, each part, each organ or function having its own in a perpetual interaction which constitute an ensemble or a whole' (Lefebvre, 1996: 230). Different global flows of ethnic groups, finance, media, and ideologies have disrupted the search for coherence or unity by local economics, culture and politics (Julier, 2000; see also Appadurai, 1990). Instead what has occurred is a particular hybridisation where these global flows mix with and in the local realm to generate new place identities. The linear tunes of developers have been remixed by a variety of local and global forces into a unique combination of tempos, intensities and tunes that produce El Raval's distinct public life.

Inevitably, the story is not straightforward and two complications must be taken into consideration. Firstly, as Lefebvre argues 'Polyrhythmy always results from a contradiction and also a resistance to it – of resistance to a relation of force and eventual conflict.' (1996: 239) Thus, while rhythms partially co-exist in El Raval, this has not been without initial conflict. New sensescapes code the cultural meanings of places differently and many of El Raval's established residents have been feeling threatened and overruled by the new sensescapes developed both by immigration and tourism. Hence, over the last ten years several incidents of protest against tourism and immigration have occurred. The underlying discourse is frequently based on sensuous paradigms such as protest against the noise of bars and restaurants, the smells left by urinating night-revellers, or the perceived lack of basic sanitary conditions in immigrant homes.

Secondly, the emergence of the marketing strategy 'ravalejar' in 2005 is indicative of a new phase of place promotion and change in processes of commodification in the city. The campaign was conceived by the *Fundació Tot Raval*, a platform set up in 2001 by a mixture of cultural entities, official authorities, private businesses and individuals from the neighbourhood whose aim is 'to act as an intermediary with the authorities and to sensitise citizens to the positive image of Raval and erase forever the negative image that had been created' (see www.totraval.org). Financed by Barcelona's city council, it was conceived as a communication campaign to promote positive values about the neighbourhood. As Tot Raval states: 'El Raval is more than a neighbourhood, it's an attitude, a way of doing and living. This is why the campaign was based upon the conjugation of a verb, so that it would not be static, so that it is alive and that everybody can create his/her own version of ravalejar' (Fundacio Tot Raval, 2005). The campaign consisted in making this verb appear in a variety of spaces and 'elements of everyday life' (ibid.) in El Raval: as a slogan on the walls around the Museum of Modern Art and other parts of the neighbourhood, and as a mobile object on t-shirts, posters, bags, lollipops and table mats. Furthermore more than 23 restaurants offered menus, cocktails or other products called 'ravalejar'. Fundacio Tot Raval has described the campaign as a success as 'ravalejar' has gained its own momentum and is used as a term in the local (and increasingly) international press (for example *Barcelona Lonely Planet Guide*, 2006).

There are various issues to take into account here. On the one hand we could understand the campaign in Lefebvre's terms as a commodification of lived space by conceived space and capital. After the council's unsuccessful attempts to control and manipulate the sensory rhythms and public life of El Raval with the regeneration of the neighbourhood, this campaign recycles and appropriates the urban rhythms that already exist. The campaign does not aim to control public life but appropriates it under a coherent narrative with the intention to change perceptions and attract more visitors and consumers into the neighbourhood with a clear economic aim. Following Pine and Gilmore (1999) 'ravalejar' can be seen as an expansion of 'experiential marketing' where '[e]xperiences represent an existing but previously inarticulate genre of economic output' (ix). Experiential marketing is adapted and applied in the entrepreneurial city in the promotion, and thereby, consumption of public space. Indeed, 'ravalejar' infers following a specific spatial route as the campaign is supported mainly by El Raval's hip new shops and cultural establishments located in the regeneration corridors. The slogan promotes a particular spatial practice and engineers circumscribed forms of experiencing the diversity of rhythms in the neighbourhood. It entails a selective process as to which aspects our senses are to be alerted. Specific configurations of sensory rhythms configure into distinct sensescapes where El Raval's designer bars; cutting edge shops and boutiques mix with immigrant lifestyles and the grittier aspects of the neighbourhood. As one does 'ravalejar' a distinctive somatic landscape emerges as the consumer is invited to participate in and be the producer of a lifestyle: 'theming an experience means scripting a story that would seems incomplete without the guests' participation' (Pine and Gimore, 1999: 48).

The success of the marketing campaign 'ravalejar' highlights the desire by city users for less homogenised and more polyrhythmic environments in which a multiplicity and diversity of sensations are possible. It also draws attention to a new trend in contemporary urban planning and marketing campaigns to frame and regulate urban rhythms so that 'certain kinds of multiculture become visible [and sensible] because of the visual [and sense-able] ordering of the spatial' (Keith, 2005: 175). So, while El Raval's ethnic and cultural mix has repositioned this once unmarketable area as one of the trendiest neighbourhoods in Barcelona and is attracting more than 25 million visitors a year (*El Periodico*, 27 May 2008), El Raval's poverty and health indicators remain one of the lowest in Barcelona (Subirats and Rius, 2005).

Conclusion

In the case of El Raval we can clearly identify a transformation in processes of commodification of space where visual consumption, the 'tourist gaze' (Urry, 1990), is expanded into a more holistic consumption of sensory rhythms: textures, sounds, smells and even tastes. Howes (2005) describes this increased commercialisation as the sensual logic of late capitalism. In his view, as consumer landscapes have become more alike and visual fatigue quickly sets in, touch (and sound, taste and smells) revivify. As I have shown, the campaign 'ravalejar' implies not only the commodification of architectural or urban spaces but the sensorial consumption of the lifeworlds of residents and visitors, who through their spatial practices define the atmosphere of the neighbourhood. From a cynical point of view one could argue that marginality, often related to unexpected and uncontrollable experiences, is becoming a desired attraction in increasingly homogenised cityscapes. The existence of immigrants, of prostitution, of extreme poverty are 'signs of authenticity' for a neighbourhood (Zukin, 2008), as long as they are interspersed with trendy bars, restaurants or entertainment and can be left behind once the visitor can step into a chic venue or close the doors of the newly regenerated loft apartments.

The explicit focus of this chapter on urban rhythms illustrates changes in the ways that urban regeneration and place branding operate. It highlights how an important strategy is to transform the sensory-experiential geography of places. As bohemian gentrifiers and 'cool' tourists shun generic and commercialised spaces in search of places outside mass consumption, city councils and urban marketing professionals are consciously searching for, producing, managing and commodifying novel urban rhythms in edgy and often marginal neighbourhoods from Barcelona to Paris, Tokyo to São Paolo. Yet, while there is a wish to fix and manipulate urban rhythms the example of El Raval has shown that these are fluid, elusive and slippery – constantly changing through cultural adaptation, resistance and mutation.

I want to thank Germa Iturrate from the Arxiu de Ciutat Vella for helping me to research the campaign 'ravalejar'.

Chapter 3

Life Hacking and Everyday Rhythm

Tracey Potts

> It is time to write The Death of the Clinic. The clinic's methods required bodies and works; we have texts and surfaces. Our dominations don't work by medicalization and normalization anymore; they work by networking, communications redesign, stress management. (Haraway, 2004: 30)

Information management is becoming big business. As information proliferates and becomes increasingly informated (Liu, 2004), further information is produced around questions of dealing with such informated information. What becomes clear is that living successfully in the 'integrated circuit' (Haraway, 2004: 7) demands new strategies and tactics, new attitudes – what Alan Liu (2004) terms new *workstyles* – new ways of dealing with data in all of its 'material-semiotic' specificity (Haraway 2004: 201). The shift to digital inaugurates the creative destruction of Fordist and Taylorist blueprints and ways of operating (Liu 2004). New recommendations for the division of labour – coded as personal productivity tips or 'life hacks' (a term coined in 2004 by Danny O'Brien to describe the operations of Silicon Valley programmers in dealing with information overload) and branded as David Allen's *Getting Things Done®* system (2003) – follow in the wake of such destructions. In this chapter I consider these recommendations rhythmanalytically, drawing on Henri Lefebvre's (2004) claim that dominance is secured through rhythmic intervention in the everyday. The resulting portrait, certainly, is gloomy but in bringing the material, energetic aspect of information to the fore I hope to allow a strain of immanent critique (that is, everday mundane resistance) to emerge. Focusing upon the procrastinating body as a finely calibrated instrument of contemporary digital rhythm – in Lefebvre's terms, as a metronome – introduces an element of static into the communication circuitry, static which can be figured as a productive brake upon the strategic imaginary of information as continuous, unimpeded *flow*.

Livebetterworkbetter.com

If cultural intermediaries help instruct us with lifestyle questions – what to (and not to) wear, eat and buy – then the equivalent in the business world might be termed corporate intermediaries: the question of how to (and not to) work manifests in an expanding universe of popular self-help business literature. Aside from Allen's GTD®, Neil Fiore's *The Now Habit* (2007), Stephen Covey's *The 7 Habits of Highly Effective People* (1990), Mihaly Csikszentmihalyi's *Good Business: Leadership, Flow, and the Making of Meaning* (2004), and Patricia Ryan Madson's *Improv Wisdom: Don't Prepare, Just Show Up* (2005) offer strategic systems for dealing with information overload and its effects: stress, burn-out, procrastination and generalised inefficiency. Not merely a cure for office ills, however, personal productivity literature promises boundless gain: if Fiore promises leisure free from guilt, Allen, the professed 'Henry Ford of the digital age', assures us that his system will provide 'a better way to work and live' (GTDtv, 2009). Ryan Madson, upping the ante, offers 'guidance for a life of adventure and meaning' (2005a). Self esteem, personal and spiritual growth are all, consequently, at stake in learning to work more effectively.

Strategic systems are, in turn, themselves converted into useable utilities and ground level practices and constitute a formidable presence in the blogosphere known as lifehacking (Blogcatalogue.com lists 9,556 items for personal productivity and 19,527 for lifehacking). The David Allen of the on-line lifehacking community, Merlin Mann, runs a successful website – 43 Folders. com – which offers tactical advice on implementing GTD® in particular. Other sites specialise in offering creative solutions – keyboard shortcuts and 'killer apps' – to digital problems such as synchronising files, tracking tasks and managing email. Lifehacking is continuously becoming more targeted in focus; in addition to general sites such as Lifehack.org (lifehack wiki), Lifehack.net, lifehacker.com, Productivity501.com, zenhabits.net, lifeclever.com, the cranking widgets blog, hacktheday.com, cleverdude.com and Wisegeek, specialised hack advice can be accessed at Parenthack.com, Hackcollege and academic productivity.com.

If lifehacking begins as a 'techy' office phenomenon, it soon becomes dispersed toward what lifehackers refer to as life 'off-line'. Recent posts on Lifehack.org included details of a software package to keep family relationships intact and weight loss guidance for a perfect Valentine's Day, how to plump up flattened sofa cushions in the sun, how to fold towels so that they look like animals (for some Friday fun!) and 34 tips for your younger self. Lifehacker.com offers advice on childproofing your home theatre system, decluttering (with the use of a '6 month maybe box'), how to remove permanent marker from walls, manage a to-do list with your iPod Touch, avoid an IRS audit, build a Nerdcycle (a customised exercise bike enabling the user to simultaneously surf the internet and keep fit) and make conferences worth the trip and time.

Despite the prevalence of iPods and software widgets, 'alpha geeks' (O' Brien, 2004) are by no means 'over' analogue technologies; there is a resolutely low-tech movement within the lifehacking community which favours paper and pens, albeit refracted through geek speak. The hipster pda, for example (which comprises index cards and a bulldog clip) is constituted as an 'analogue data capture device' (Mann, 2004). Douglas Johnston's D*I*Y planner (2006) offers a free downloadable set of templates that can be printed directly onto index cards, while PocketMod (as recommended by Kim Komando – America's Digital Goddess™) turns a single sheet of A4 into a personal organiser (PocketMod, 2007). For those afflicted with notebook fetishism there is a world of tips for hacking Moleskine's into flexible utilities including an entire page devoted to Moleskine friendly fountain pens or 'data capture tools' (43 Folders, 2007). Life hacking is, then, less about the actual technologies than the manner or style of using them. To summarise: if this is the short story of lifehacking, the haiku might be, in the words of Wikipedia: 'anything that solves an everyday problem in a clever or non-obvious way' (2008).

@@@@@@@@@@@@@@@@@@@@@@

At a tactical level, the ingenuity of life hacking is more than evident; life hackers operate skillfully and inventively, moderating and adapting tips and schemes, offering an almost textbook definition of Michel de Certeau's idea of everyday, ground-level creativity:

> Tactics are procedures that gain validity in relation to the pertinence they lend to time – to the circumstances which the precise instant of an intervention transforms into a favorable situation, to the rapidity of the movements that change the organization of a space, to the relations among successive moments in an action, to the possible intersections of durations and heterogeneous rhythms, etc. (2002: 38)

As a series of wily adjustments to strategic arrangements – that we can, for convenience, code as 'demand culture' – life hacking has a good deal going for it. The pleasure of solving everyday problems creatively and skillfully is not to be underestimated (I can offer plenty of personal testimony here as a user of lifehack sites). However, I wish to consider the corporate blueprint (or *design for working*) aside from its tactical adoption in order to better appreciate its macro recommendations. Developing business projections before factoring in the disruptive creativity of social practice is crucial in order to consider what is at stake in new workstyle arrangements. The creative inventiveness of lifehacking notwithstanding, together with its unpredictable take-up (how exactly is it used, appropriated, acted on, attempted, ignored, partly adhered to, started but not continued, held as an intention), there is, discernable at the level of corporate strategy, I argue, a vision of a totalising existence that can be summarised in what Maurizio Lazzarato (2004) sees as the move from capital-labour to capital-life: the incorporation of lifestyle into systems of capital exchange. Twentieth century programmes of rational recreation and hobbyist leisure pale in comparison to their post-industrial equivalents, which, in Liu's (2004) view, offer no constitutive outside; intracultures (compensatory pleasures of the cubicle emblematised in the 'cool' screensaver and the clever widget) and their resultant microleisure activities constitute the order of the day, and weekend or 9-5 rhythms feature as obsolescence. In the era of smart work, smart workers know that the total sum of their portfolio (worklife and lifelife *flow*) secures advantage in, to borrow Allen's GTD® mantra, 'the game of work and the business of life' (2008). Shuttling between macro and micro perspectives – and importantly, beginning macrologically – allows global forces to be appreciated and everyday resistances to be assessed outside of the enthusiasms of 'early adopters' of lifehack tips.

Viewed *rhythmanalytically*, a significant scene begins to form featuring fundamental shifts in the fabric of everyday arrangements around work and leisure. Such shifts in the received idea of twentieth century division of labour make it hard not to be nostalgic for what are fast becoming obsolete timetables; to borrow Haraway's words: '(M)odern production seems like a dream of cyborg colonization of work, a dream that makes the nightmare of Taylorism seem idyllic' (2004: 8). That said, it is crucial to consider these emergent 'technological forms of life' (Lash, 2001) seriously and *non-nostalgically* (Haraway, 1997). Operating between technophilia and technophobia, between the personas of the enthusiastic early adopter and the Luddite who harks back to the era before we were wounded by machines, is, of course, easier said than done and can never be an evenly distributed performance. To reiterate, it seems necessary to begin pessimistically before celebrating the ingenuities of specific lifehacks in order to better glimpse the dynamics of the proposed new merger between labour and life.

£$£$£$£$£$£$£$£$£$£$£$£$£$£$£

With such pessimism in mind, a shift of vantage point is helpful: the domestic interior can offer a unique view of the world of work in the era of information and is recording some of the rhythmic and material shocks of that world in its very fabric. In many ways, this exploration of lifehacking is a conjoined project and it began with an investigation into a recent lifestyle phenomenon: decluttering. The interior *designs for living* issuing forth from lifestyle TV and magazines seemed obsessed with storage and with the idea that our things were getting in the way of successful living. Lifestyle professionals – variously figured as consultants, doctors and gurus – made grand claims for the practice of tidying and rationally ordering the object world. Removing the perceived energy blockages manifested in clutter became the key to enhance energy flow, cure illness, improve productivity and relationships, relieve stress and add value to property in the process (figured as a bonus in much of the advice, the explicit agenda of monetary profit is mostly euphemised, not to say denied). Unsurprisingly, redescribing these claims sociologically revealed the flow of energy to be a guise for the flow of capital. Decluttering turns out to be utterly concerned with money: with producing more efficient consumers – freeing up space and converting things in the attic back into cash so we can buy more things – and with manufacturing symbolic revenue through taste actions – orchestrating spacious minimalist interiors which translates into profit in situations where space is itself at a premium (Potts, 2008).

In the process of unmasking the agendas of professional lifestylers, it became clear that there were still greater forces at work and that another space for concern was starting to open up: the office. Professionals were borrowing techniques directly from the office and advocating what Wendy Wheeler (1999: 103) has termed a 'managerial approach to living', featuring hourly, daily, weekly, monthly and yearly timetables, filing systems, labels, colour codes, databases, card indexes, dedicated storage systems, action trays, to-do lists, memo boards, targets and deadlines. More, the division between home and office – especially in the form noted by Benjamin (2002) where the domestic interior is figured through a restorative softness in relation to the harsh world of work – seemed to be dissolving. The prevalence of work space in the home – demarcated in normative schemes as the home office, together with living solutions that established continuities between living and working – demonstrated that for certain among us – white collar professional classes, knowledge workers – moves were afoot to shift the furniture schemes of capitalist modernity and with them traditional divisions of labour.

Over in the office complementary cultural shifts could be detected. As TV makeover queens Dawna Walter (2002) and Ann Maurice (2001) were castigating us for our slovenly domestic habits, IBM publicised the appointment of a Storage Guru. Clutter was not only a domestic issue, 'the paper beast' was stalking the office and office constipation was infecting workspace (Aslett, 2005). No matter that some academics were seeing meaningful personal display in the objects that accrued in cubicle and on the desktop (Wells 2000), others saw obstructions to productivity. More, if the domestic realm appeared to be undergoing professionalisation, then the world of work – of smart work, especially – was embracing play and creativity whilst directing its energies to the extra-curricula. Inherited divisions and between home and office and alignments between, on the one hand, home/leisure/play and, on the other, office/work/business were being reconfigured.

In the context of twentieth century work/leisure programmes, such relaxations seem wholly welcome, especially in relation to Adorno's portrayal of the psychodynamics of 'free time': as not only commodified, 'shackled to its opposite' (2001: 187) and hence unfree, but structured according to a bi-polar and neurotic relation between punitive parent and undisciplined child. The idea of compensatory leisure and schemes of deferred gratification, organised around notions of clocking on and off, boredom and restorative excitement, drudgery and reward, acknowledges tacitly the fact that worklife, for the most part, is something to be endured. The promise of an integrated existence based upon 'relaxed productivity' (Allen, 2008) together with the potential for a creative and stimulating workplace appears progressive, adult and healthy.

Microrhythmia

Looking at the shifts in operating at home and at work rhythmanalytically, however, allows a different picture to emerge, especially in relation to the retraining of the body that is a constitutive feature of lifehacking. Lifestyles and workstyles are improved through alterations in the basic tempo of existence. Efficiencies are often achieved through modifications to the daily timetable. Many hacks start with tips about task frequency and recommend a micromanagement approach. Attending to what Lefebvre determines as the 'lived temporality' (2004: 6-7) of such schemes requires the acknowledgement of the centrality of rhythm:

> The commodity prevails over everything. (Social) space and (social) time, dominated by exchanges, become the time and space of markets; although not being things but including rhythms, they enter products. The everyday establishes itself, creating hourly demands, systems of transport, in short, its repetitive organization. (ibid.)

More, everyday rhythm bifurcates into cross-cutting, competing pulses and repetitions: the cyclical, cosmic motions of 'days, nights, seasons, the waves and tides of the sea, monthly cycles, etc.' (8) intervene with and are compromised by the 'linear repetitive', the time of social practice. For Lefebvre, the establishment of a commanding linear rhythm is the key to establishing dominance:

> Objectively, for there to be change, a social group, a class or a caste must intervene by imprinting a rhythm on an era, be it through force or in an insinuating manner. (14)

If, in our time, the 'linear tick-tock', the time of social practice, is moving from the mechanical temporal progression of commodity capitalism to that of the digital clock and information and knowledge exchange, tracing new patterns of organisation becomes politically charged.

Like certain other animals – horses and dogs in particular – modern humans require breaking in (40) and are hence, as Foucault also argues, subject to rigorous disciplinary training. Lefebvre's notion of *dressage* brings Foucaultian ideas into the realm of the everyday, however, establishing how the social world imposes its structures, laws and rationalities whilst honouring the quivering, vital pulses of the body – 'respiration, the heart, hunger, thirst' (8). The cosmic may be overwritten by the linear repetitive but the body has the last word. Hearts beating, hands trembling, breath quickening 'inaugurates the return of the everyday' (ibid.). Repetitive organisation is ensured only a partial victory even in the most well trained workhorses. As Lefebvre notes, '(A)bsolute repetition belong to maths and abstract logic' (7), for 'in practice and in culture, exhaustion is visible sooner and more clearly than growth or innovation' (15). If the aim of dressage is to calibrate bodily rhythm to suit the pace of social rhythm – to establish *eurhythmia*, Lefebvre's term for culturally sanctioned rhythmic operations and ways of being, determined through normative conceptions of health and well-being – then the prospect of mis-matched, contrapuntal beats poses a permanent threat. *Arrhythmia*, 'fatal desynchronisation' (ibid.) at its most extreme, between body and cultural body, constituting failure in normative terms, receives one of its most famous portraits in Charlie Chaplin's *Modern Times*. As a potential slave to twentieth century machinic rhythm, Chaplin's character fails in spectacular style. Viewed rhythmanalytically, *Modern Times* offers up a polyphonic display of bodily resistance and immanent critique.

Attending to the whispers and gasps of the cosmic body in post-industrial times requires more subtle instruments though. Lefebvre characterises the cultural shifts of modernity thus:

> Symbolically (so-called modern) *society* underwent something that recalls the great changes in communications. It saw the cylinders, pistons and steam jets on steam engines; it saw the machine start up, pull, work and move. Electric locomotives only present to the eye a big box that contains and conceals the machinery. One sees them start up, pull and move forward but how? The electrical wire and the pole that runs alongside it say nothing about the energy that they transmit. In order to understand, one must be an engineer, a specialist, and know the vocabulary, the concepts, the calculations ... (15)

Becoming an engineer of the electric age is one thing, but how are we to gauge a world without poles and wires – a digital age, where as Haraway notes, 'our best machines are made of sunshine' (2004: 12); where there is no big box or material means of concealment? In considering what remains visible in the virtual world Lefebvre falls back on the everyday. With this, bodies as registers of traces, as processors of demand, bodies that tire, over-compensate, collapse, recover, go slow, become figured as *metronomes* (15), as the means by which dominant rhythms becomes legible. What then can the procrastinating body – and more, the life hacker's proposed remedy – tell us about demand culture?

Back in the office, a cure is being offered for the pandemic of the information age: procrastination. Procrastination is to the office what clutter is to the domestic interior and its treatment follows a similar course. Fostering a Now Habit requires the identification of the particular psychodramas that cause one to hold onto energy (Fiore, 2007). The professed 'thief of time' (according to a mousemat available from amazon.co.uk, R.R.P. £4.99), procrastination finds a cure in the temporal reorganisation of the working day. Prescriptions include Francesco Cirillo's (2007) Pomodoro Technique (named after a tomato-shaped kitchen timer that structures work and non-work time), Fiore's 'unschedule' (2008), and Merlin Mann's (2005) lifehacked version: the (10 plus 2 times 5) Procrastination Hack which works through short bursts of focus punctuated with shorter breaks. A timer is set for 10 minutes, during which time tasks are 'bust through', when the buzzer goes off a 2 minute break is set, during which time permission is given, in Mann's words to 'dick around', when the buzzer goes off a further ten minute period is set and so on until an hour has passed and 50 minutes focused productivity has been achieved with 10 minutes of 'reward'.

From Lefebvre's point of view, this is dressage of a kind that speaks directly to our propinquity with those animals who offer use-value:

> One breaks in another human living being by making them repeat a certain act,
> a certain gesture or movement. Horses, dogs are broken in through repetitions,
> though it is necessary to give them rewards. (2004: 39)

While some bloggers are suspicious of the buzzer technique (Noteberg 2008),
many lifehackers maintain that micro work/reward structures respond to the body's
natural rhythms (in Lefebvre's terms are hence cosmically eurythmic). Mann's call
to 'honor thy energy' chimes with Cirillo's claim to dismantle Taylorist blueprints:
'with the Pomodoro Technique, there is no inspector who monitors workers' hours
and methods in a Tayloristic fashion' (2007: 32).

The idea of allowing the cosmic to determine linear time is contradicted by
much lifehacking advice, however, which maintains the importance of *building* a
habit, precisely by implementing a micro rhythm of work practice. 'Make every
moment productive' suggests one lifehack blog (Anon, 2007); approach work, and
by extension life, as a body builder or athlete might, 'oscillating between carefully
orchestrated periods of rest and activity… your days and months should be broken
down into smaller units of cycling between productivity and rest'. 'Discipline is
a muscle' and 'habits are what make things easier' so 'setting an hourly goal for
what you want to accomplish with a strict deadline can push you to action' (Anon,
2008). Turned into sociology, lifehackers begin by hacking the habitus (Bourdieu,
1986). Far from the cosmic leading the linear, the demands of demand culture set
the pace for tighter and tighter circuits of oscillation.

The author of *Improv Wisdom* further betrays which kind of time has the upper
hand in recommending 'yes' as the most important word in the smart workers
vocabulary. Literary Modernists will be interested to note that the model of 21st
century productivity comes in the shape of Molly Bloom:

> This is going to sound crazy. Say yes to everything. Accept all offers. Go along
> with the plan. Support someone else's dream. Say "yes"; "right"; "sure"; "I
> will"; "okay"; "of course"; "YES!" Cultivate all the ways you can imagine to
> express affirmation. When the answer to all questions is yes, you enter a new
> world, a world of action, possibility, and adventure. Molly Bloom's famous line
> from *Ulysses* draws us into her ecstasy. (…) Don't confuse this with being a
> "yes-man," implying mindless pandering. Saying yes is an act of courage and
> optimism; it allows you to share control. It is a way to make your partner happy.
> Yes expands your world. (Madson 2005: 1)

Saying yes, opening oneself fully, ecstatically – mmmm yes! – to the demand
offers protection from the cardinal sin of blocking:

> Blocking comes in many forms; it is a way of trying to control the situation
> instead of accepting it. We block when we say no, when we have a better idea,
> when we change the subject, when we correct the speaker, when we fail to listen,

or when we simply ignore the situation. The critic in us wakes up and runs the show. (ibid.)

The idea that the 'yes muscle', like the discipline muscle, is in need of exercise and training offers tacit admission of, at the very least, a *contest* between the cosmic and the linear. Rather than honouring our energies, the practices of improv wisdom teach us to pathologise the messages that our bodies might be trying to pass on through allergens in the communications immune system: stress, repetitive strain injury, attention deficit disorder, procrastination and the myriad ways in which the body says 'no'.

The smart worker – mind like water, martial arts trained, saying yes! Yes! – likewise, maintains the flow of the office system by establishing seamless connectivity and indetectable modal change (which in the language of the everyday includes emailing the office in one's pyjamas without it showing). Absorbing the shocks of linear time, smart workers present zero resistance in the production circuitry, allowing the current of demand to pass straight through the body. Arrhythmic disturbance is pathologised, individualised as a personal issue, as fear of failure, fear of success (which, according to Fiore is fear of *future* failure), self-sabotage, and perversity (allowing the critic to ruin the show).

Similarly, the reduction of reward to what Haraway, in her work on dogs, would see as a *click and treat* structure (2003), installs a system of microleisure and relaxation within a hegemony of 'relaxed productivity' (which turns out to be *always on* productivity). Liu distils the situation of the post-industrial information worker compellingly:

> The granularity of our leisure … is now set not just to the scale of our evenings, weekends or annual vacations, or even to that of our lunch breaks and fifteen minute work breaks. Rather, just as the total tonnage of the world's bacteria exceeds that of all other creatures by orders of magnitude, so the leisure that now bulks largest is microscopic. The good life is microleisure – for example, the ergonomic chair that injects small doses of comfort throughout the day, the rounded corners and muted colors of a cubicle evoking something like a transient rest home for the chronically overworked; the advanced photocopy machine that collates and staples automatically so that we might have a moment of brief, vacuous respite staring into space; and above all, the computer interface whose ever more technically "sweet," "neat," and … "cool" ways of saving a step here and preventing a mistake there blur the line between ease and use. (2004: 163-4)

Killer apps and keyboard shortcuts are the order of the postindustrial day; micro and nano compensations for increased productivity (which is, like virtue, its own reward) establish digital eurhythmia.

If the Frankfurt School detected the rhythm of the conveyor belt in the shocks and starts of culture at large, then the pulses of contemporary machines – optical cables and wireless networks – present, potentially, a less *discernible* modernity. Liu cites Castells in drawing attention to the bandwidth of advanced capitalisms' processes of technological and other progress: '(T)he spirit of informationalism is the culture of "creative destruction" accelerated to the speed of the optoelectronic circuits that process its signals' (2). Following this, the microswitchings of smart work – of a cool and agile workstyle – take their lead from the zeros and ones of binary code; on/off, work/don't work oscillating in imperceptible rhythm, a harmonious flow of productivity. The big on-offs of industrial modernity – evenings, weekends, bank holidays, factory shutdown, potter's and other worker's holidays (Holliday and Jayne 2000) – dissolve in the liquid age into permanent globalised flow. Certainly, Zygmunt Bauman's (2000) image of a *liquid* modernity is deeply suggestive in the context of the current emphasis on flow: of information, capital, workstyle and lifestyle.

Microleisure, with this implosion of reward into work time and space, destroys the possibility of work's '*recreational outside*' (Liu, 2004: 77, my emphasis), replacing what in the twentieth century were referred to as subcultures and countercultures with intracultures. Liu sees contemporary manifestations and practices of 'cool workstyle' as intracultural activity par excellence:

> Cool is an attitude or pose from within the belly of the beast, an effort to make one's very model of inhabiting a cubicle express what in the 1960s would have been an "alternative lifestyle" but now in the postindustrial 2000s is an alternative workstyle. We work here but we're cool. Out of all the technologies and techniques that rule our days and nights, we create a style of work that is "us" in this place that knows us only as part of our team. Forget this cubicle; just look at this cool Web page. (78)

The vacuousness of the 'dick around' – diminishes the material possibilities of the break from work to the point that even pastimes that Adorno would have disdained as mindless become impossible. (It is tempting to speculate that the philosopher who shunned hobbies would be, these days, nostalgic for the time and material engagement summoned by the culture industry's commoditised leisure products). Non-work folds into intracultural style: '(K)nowledge workers are never far from the cubicle, where only the style of their work lets them dream they are any more than they know' (78). Cool workstyle, further, confuses the boundary between strategy and tactics; in Liu's words, 'how do knowledge workers act up in their cubicle(s)?'(ibid.), when permission is granted by strategists to 'dick-around'? What are the possibilities for the *peruque* – de Certeau's (2002) codeword for the artful appropriation of company resources (time, paper, paperclips) – within the intraculture of knowledge worker cool? (If I can be permitted a moment's self-absorption: how do I know if my vintage silver Yard-O-Led pencil, my 1940s Parker Vacumatic, my J. Herbin Café des Isles ink, together with my plain white index cards held together with a bulldog clip are doing more than injecting a frisson of intracultural steampunk style or nerd-chic into my workspace?)

In rhythmanalytic terms it is hard not to see lifehacking as post-millennial dressage, a bodily and psychic retraining for liquid modernity/demand culture and hence as victory by insinuation. The sincerity of the promises of relaxed productivity and the revenues of adventurous workstyles are offset by the medicalisation and pathologisation of what become constituted as failures to adapt to the new rhythms of demand. Apart from offering strong evidence that Haraway's call for the death of the clinic, even twenty or so years on, is premature, the portrayal of procrastination as individual (and stubborn or perverse) resistance, contradicts the presentation of personal productivity as being within human, rather than business, interest. The urge to honour one's energy seems disingenuous, to say the least.

Conclusion

Less gloomily, the prevalence of hacked GTD systems and hacks of hacks illuminates an abundance of tactical intervention; on the ground, creative inventiveness. More, the sheer volume of hacks of hacks of hacks, random advice and 'blogage' generates further *immanent critique* in that information designed to deal with information can add to rather than cut through the pile of stuff mounting on the desktop. In this respect, lifehacking is ironically a clutter generator, as a podcast detailing how to listen to podcasts in half the time betrays. Similarly, Zenhabits – orientalised Buddhism hacked for the 21st-century business world – offers sage guidance for cutting through to what's important: *72* ways to simplify your life (Anon, 2007a). Lifehacking promises us the answer to all our problems, if only we had the time. (Revealingly, David Allen stopped blogging in 2006: 'I'd probably continue it in some form, if I didn't have a multitude of other things to do that are taking priority' (2006); while Mann's recent posts berate the excesses of GTD culture, shifting the emphasis of 43 Folders to creative work over productivity (2008).

In tracing new arrangements in the spheres of private and public, domestic interior and office; in moving between lifestyle and workstyle, material culture and information overload, clutter and procrastination, profound alterations to the rhythmic order of life seem to be being orchestrated. What is of major concern here is the way that the blueprints of such alterations – designs for living and for working – place the burden of responsibility upon the individual; the emphasis on personal psychodrama is evident in both lifestyle and workstyle advice. Further, the erosion of the recreational *beyond* together with the generalised appropriation of the creative sphere to an extension of the curriculum vitae marks the inauguration, to recall Lazzarato (2004), of capital life, a situation that simultaneously inaugurates a class of those unable to keep up and to go with the flow. If Liu (2004) notes the relegation of what he terms 'matter work' to the developing world then Bauman offers a comparable scene in the image of 21st-century information age poverty: the situation of being stuck with obsolete material possessions, that is overburdened with materiality and, crucially, with history. Matter and history figure in the circuit diagram of digital progress as impedance – stuck energy (past 'issues' interfering with the present situation) and congealed possessions (sentimental attachment to things) – blockages, which must be eliminated. As pessimistic as this may appear, de Certeau's and Lefebvre's faith in the everyday to generate resistance abides and points in the decisive direction of ethnographic work; the *tactical* aspects of lifehacking together with the notion of immanent critique – the intransigent *materiality*, the stuffiness of information – present the possibility of alternative arrrangements to the rhythmic ambitions of corporate intermediaries.

Chapter 4

The Aesthetics of Place-temporality in Everyday Urban Space: The Case of Fitzroy Square

Filipa Matos Wunderlich

Introduction: a sense of time in urban space

As we travel and change location, we may notice that some cities are characterised by a vivid and contrasting sense of time, and at a smaller scale; so are particular places within cities. Such cities and places are commonly perceived as fast or slow. Fast cities may be represented as complex, busy and agitated, and their everyday social life portrayed as repetitive, accelerated and homogenised. In contrast, slow cities might be conceived as quiet and ordered, their everyday life patterned and distinct. Specific urban places within cities are also perceived as temporally distinct. Some form hectic hubs of activity and movement, where everyday life is performed in a speedy and anxious manner. In other places, social activities and movements intertwine harmoniously and are performed in a slow and relaxed fashion. These slower places are often experienced as temporary halts in the city, breathing occasions, moments of silence and encounter.

These common temporal experiences suggest a sense of time as not only somewhat *intersubjective* but also *place-specific*. Accordingly, time in urban places is produced and perceived jointly. As people perform their tasks in everyday life, they perform time collectively. Moreover, as everyday urban life accelerates and home – work distances increase and affect personal and social times, time increasingly becomes a conscious and collective object of concern.

This chapter explores a sense of time insofar as it relates to the specifics of urban places, and as it varies according to the distinctive aesthetics associated with temporality. Firstly, this temporal aesthetics is *sensual*, often associated with a sense of pace, whether perceived as slow or fast, static or dynamic, continuous or intermittent. Secondly, it is characterised by a particular sense of flow and a soundscape. Thirdly, this temporal aesthetics is *affective*, often perceived as calm and harmonious or stressed and agitated, or as either social or intimate. Finally, it is *rhythmically expressive*, structured and represented by the particular everyday rhythms of place. Places are temporal milieus within which repetitive everyday activities, spatial patterns and cycles of nature interweave and orchestrate into bundles of expressive rhythms. These rhythms are unique to particular locations, and structure and affect a sense of time in unique ways, shaping distinctive forms of *place-temporality*.

This chapter is thus dedicated to the temporal aesthetics of urban places and the notion of place-temporality in the way it is defined through its experiential attributes as a means to advance the understanding of a sense of time in urban space.

Place-temporality

Place temporality is a sense of time that is place-specific, unique to specific locations, and intersubjective, practised and perceived collectively (Wunderlich, 2008), as discussed above. Place-temporality can be explored by focusing on its sensuality, performativity and aesthetic significance, or it can be identified as a process or sequence of unfolding processes by focusing on its structure and the rhythmic patterns that constitute it.

Place-temporality can be conceived as a form of representation of time in urban space, and as such its structure is sensuous and expressive (or representational) (Duffrene, 1973, in Silverman, 1975: 464). Its sensuous qualities are revealed by dynamic and resonant corporeal gestures, bodily movements and other complex articulations. It is represented in that a sensual rhythmic structure is evident but depends upon *reflection* to be fully perceived, an expressive object that can be consciously performed and observed, and may incite care, enjoyment and imagination.

Place-temporality, however, is also defined as a sense of time that results from immediate sensual experiences and as such, involves *feeling*, a 'certain affective quality' (ibid.). This is not mere emotion, but a meaningful (both sensual and affective) appreciation that defines our relationship to place. It is thus not an intellectual understanding of time, but a distinctive temporal experience of place that is sensually valued and affectively remembered; an aesthetic experience.

In what follows, I explore the aesthetic dimension of place-temporality, first by focusing on the nature of its aesthetics and second by focusing on its particular experiential attributes. The latter is illustrated through the case of Fitzroy Square in London.

The aesthetics of place-temporality

The aesthetics of *place-temporality* is bound to experience and the process of perception, and in this way its meaning is close to the original meaning of the word *aesthetics,* employed in eighteenth century philosophy to designate a 'level of cognition that one receives from immediate sense experience prior to the intellectual abstraction which organises general knowledge' (Korsmeyer, 2004, quoted by Bowman, 2006). This definition is closely related to the original meaning of the Greek word *aesthesis*, meaning *feeling* or *sensation*, and the word *aisthestai* meaning *to perceive*. This sense of aesthetics thus refers 'to more than just the beautiful and the sublime' (Boenisch, 2007), contrasting with a modernist *high art aesthetics*, a neo-Kantian interpretation that privileges the visual sense and the 'contemplative distance between audience and the artwork' (Korsmeyer, opcit).

The aesthetics of place-temporality involves a feeling and the very act of perception through which meaningful (sensual and affective) accounts of the aesthetic object (as temporal event) arises. It is *a sensual aesthetics* that recognises all senses, but principally the aural, the visual and the haptic. Thus one experiences place-temporality by being in touch with the place, listening to its sounds, and observing what goes on. It is also *an experiential aesthetics*, the aesthetic/significant object (as an event) being the very act of experience. One dwells through time in place, and therefore an aesthetics is perceived through a mode of praxis. Accordingly, it is also *a performative aesthetics* that relies on the bodily involvement of the spectator/reader as performer. Only through everyday corporeal and sensory involvements (being in place-time) is it possible to discover the temporal, sensual and affective qualities of urban place, and the aesthetics of place-temporality. As urban places comprise a multitude of complex rhythmic events that synchronise and repeat over time, with their identifiable sensual and affective (or meaningful) qualities, the aesthetics of place-temporality may be defined by the entangled temporalities of social life, nature and space. As a consequence, it can further be understood as *an everyday aesthetics*. These aesthetics qualities have been explored in phenomenological studies, particularly in the work of John Dewey, who looks at the aspects of the notion of 'art as experience', and also defines an aesthetic of routine and ordinary experience.

An art form: an experiential and performative aesthetics

Dewey (1934) 'recognises a vital connectivity between art and ordinary experience' (Highmore, 2004: 315). The aesthetics of place-temporality can be related to the particular notion of *art form*, namely *experience as art* arising from sensorial and corporeal involvement; its significance surfacing through performance and practice. Here, the significance of temporal processes derives from the interaction of the body with the environment in an immediate context. Inherent to these notions of the *art form* and *experience as art* is also the Aristotelian notion of *praxis*, the 'action the goal of which is the action itself'. Praxis is opposed to *poesis*: the 'action the goal of which is the product of that action' (Määttänen, 2005: see also Barthes, 1985). In this context, place-temporality can be understood as *an art form*, grounded in this understanding of 'praxis as life' and 'praxis as art', as its sensuality and affective significance only surfaces in the moment it is performed and through the practice and recognition of routine in everyday life, as is now further discussed.

An everyday aesthetics

Dewey also develops an aesthetics of routine composed of two different types of experience: 'ceaseless flow experiences' and 'experience events' (Highmore, 2004: 315). The first are unconsciously experienced events, continuous and repetitive everyday routines whereas the second are those experiences that stand out from the former and are subsequently vividly remembered. Although Dewey only regarded the 'experience events' as significant, Mikhail Csikszentmihalyi (1988) demonstrates the importance of flow experiences in the enjoyment of our everyday life. Both flow and event experiences combine to produce fulfilment and are thus aesthetically significant. For in urban space, both experiences constitute and represent the experiential continuum of everyday life and define the sensual and affective temporal senses of place. As such, they shape the experience of *everyday place-temporality*.

Other aspects of the experience of everyday place-temporality are similar to the experience of routine and ordinary experience. Dewey writes that the experience of routine is characterised by 'slackness, inchoateness and drift', 'slackness' involving a 'diffuse consciousness of routine'; 'inchoateness' meaning a distracted and unfinished experience of 'tumbling profusion and messiness'; and 'drift' the process of 'picking up and letting go of concentration' (Highmore, 2004: 316-317). Awareness of these attributes of ordinary experience help bring to consciousness the experience of rhythm and time (Flaherty, 1998; Lefebvre, 2004). Ordinary experiences are intrinsic and meaningful in urban space. They thicken the temporal grain of places and shape everyday place-temporalities into sensual and affective (aesthetic) experiences.

The aesthetic experience of place-temporality: the case of Fitzroy Square

I now exemplify three aesthetic attributes of the meaningful experiences of urban space that form *place-temporality*. Firstly, I investigate the aforementioned vivid sense of time, whether slow or fast, calm or agitated. Secondly, I trace the experience of *flow*. Thirdly, I investigate the distinct *soundscapes* of place. These qualities are illustrated through the case of Fitzroy Square, London.

A vivid sense of time, or tempo, with aesthetic significance

As one walks into Fitzroy Square, a sense of standing elsewhere, somewhere that is temporally distinct, surfaces almost immediately. The square is characterised by a sense of *slowness*, more specifically, a sense of *slow time*, with particular sensual-affective attributes such as calm, rest and encounter, enjoyment, and a certain sense of belonging. These attributes suggest this square as a place of aesthetic significance, both sensual and affective. Fitzroy Square stands out as both temporally distinct and a meaningful social space within its neighbourhood, as the following narratives recorded in the square suggest.

> As one enters Fitzroy square, things seem to come to a halt. They don't freeze, however. People, pigeons, dogs, nature seem to synchronise in a slow dance ... social life unfolds at a different pace, different from anywhere else, and nature imposes its cycles, its presence is appreciated. On the one hand, in a harmonious way people and nature seem to perform according to a different *tempo*, the *tempo* of Fitzroy Square. On the other hand, they seem uninterruptedly to shape and influence its pace. People hold or slow their pace as if it were a place of arrival or of such sensual significance, a visual, haptic and aural significance that seems impossible to ignore. This pace is sometimes interrupted by groups of people walking faster or joggers running, or by nature, as when sudden rain or gusts of chilly wind pushes everyone out of sight. Yet, these moments of temporal conflict don't prevail and people, nature and space again settle back into harmony soon. This pace seems contagious, as after a couple of minutes sensing the square, I start to feel calmer and start walking slower. I unconsciously become sensually involved with the rhythms of the square". (extract from the author's place-diary, on-site narrative written at 15:30pm, 20 April 2005)

> "I came out of my office and the square feels and makes me feel relaxed and calm. It is a slow space". (Louise; ethnographic quote, 12:30am, 1 June 2005)

The first extract refers to Fitzroy Square's particular sense of *slow time,* a sense that is experiential and localised; a time that is specific to its physical, social and natural orders; a *tempo* one may attentively *listen to* expressing itself as an attribute of people, nature and space. Also, it suggests that this temporal sensuality is perceived and performed through ordinary experiences and everyday life patterns. The second extract, a quote, confirms its intersubjectivity and experiential character. This sense of time and place-temporality compares to the *tempo* of a musical piece, in that it is similarly experiential, intersubjective and performative. As such, it offers temporal distinctiveness and character to Fitzroy Square.

A flow experience

In the city, people sense and practise *time* whilst they are engaged and absorbed in their everyday life practices, not thinking consciously about time but apprehending it as a *flow* of experience. A strong *sense of flow* is experienced in urban places where there is also a strong and distinct sense of time. *Flow*, as conceived by Mihalyi Csikszentmihalyi, describes a quality of temporal experience characterised by the 'experiential consciousness of order and tempo' (1988: 30), a sense that is often coincident or at least relative to that of the activity being performed. Senses of *flow* typically occur while practising daily routines. Furthermore, the experience of flow is related to a distorted sense of time that emerges as full attention, awareness and concentration is put into the activities being performed. This *sense of flow* is also inherent in forms of culture that are structured and defined by fixed coded practices and rituals. Lastly, *flow* suggests as autotelic, in other words an experience that is an end in itself and intrinsically rewarding; and as a consequence, an experience people may wish to replicate since it nurtures a sense of wellbeing. Place-temporality experienced as *flow* is characterised by a sense of order and regularity as one is immersed in everyday movement and practices, a sense of time is distorted by a sensuous and affective involvement. When people's awareness of their activities and the activities merge, consciousness of place-temporality surfaces, which may produce a fulfilling experience.

People may not just generally enjoy *being in flow*, they may particularly be drawn to places with *a typical sense of flow*, with patterns of practices and events that interweave and synchronise according to a particular tempo and sense of time. Consider places where one may seemingly lose a sense of time, immersed in place activities like strolling through a market, roaming around a town centre, high street shopping or hanging out in a local park. These places relate singular place-temporalities characterised by particular states of flow. Such place-temporality is not only experienced as flow but also initiates movement towards a state of flow.

Regular and extensive experience and observation indicates that Fitzroy Square is a place that fosters *flow experiences* characterised by a sense of temporal immersion and enjoyment coupled with a sense of protracted (prolonged) *time*. Unconscious involvement with daily practices, natural events and architectural patterns is inescapable. Everyday life events are expressed vividly. One unconsciously gazes *at* them or performs *with* them. Events arrest our eye, grab our attention or induce our bodies to follow their *tempo* and practice. At the square, one easily loses self-consciousness and aconsequently performs unreflexively and in resonance with other bodies and events. Following episodes of temporal immersion, as one returns to consciousness one often feels *time was well spent.* To this sense of temporal enjoyment is attached the sense of 'protracted time' (Flaherty, 1998), a sense of *prolonged* or *slow time*. Both this sense of temporal enjoyment and slow time defines what Csikszentmihalyi (1988) calls the *flow experience*, as discussed above.

Fitzroy Square is also characterised by *a typical sense of flow*, structured by singular flow experiences that recur and are experienced collectively throughout the day. Here, social life is performed in similar manner and at a collective pace. For example, one meets up with friends and spends identical periods of time as others also do with their friends; at midday, one sits and eats lunch at a similar pace and duration as one's neighbours, and if the sun shines one enters and sits in the garden to rest along with others; and in the garden in autumn, one lays down in the grass and looks at tree leaves quivering and falling with the wind, or gazes at the sun and the clouds and their show-hide rhythm and the shadows that result, as others do also; and as the afternoon working-shift approaches, the garden closes and everyone gets up and leaves at almost the same time; and in the morning one walks faster through the square whereas at midday, one walks more slowly. Yet, none of these practices and *tempos* are deliberately choreographed; rather, they all simply unfold in synchronised manner, and one unreflexively performs along and fits into the rhythm of the square. Such singular and recurrent experiences are representative of particular moments of the day and are just some of the multiple practices and events that interweave and unfold in synchronised manner under similar *tempo* and *pace*, together inducing an overall typical sense of *flow* at Fitzroy Square.

A vivid soundscape

Henri Lefebvre (2004) and Tim Ingold (1993; 2000) suggest that the perception of temporality in everyday environments is privileged by the aural sense. The soundscape of an urban place thus plays a major role its perception. Acoustic colourations that are clear and harmonious, and the experience of sound repetition, rhythm and resonance in particular are significant indicators of place-temporality. The soundscape and sonic experience of urban place-temporality is thus vividly shaped by the aural traces of recurrent social activities and cyclical natural events in singular locations.

Soundscape refers to 'what is perceptible as an aesthetic unit in a sound milieu' (Augoyard and Torgue, 2005: 6-7). Murray Schaffer addresses the sound texture of acoustic environments through the concepts of low-fi and hi-fi soundscape; low-fi characterised by a blurred and hazy sound environment and, in contrast, hi-fi as clear and precise. The latter particularly characterises urban place-temporality. In a hi-fi soundscape, all sound 'frequencies can be heard distinctively' as they 'overlap less frequently' and are not masked by noise or other sounds (Schaffer, 1977: 43). Also, more sound perspective features, whereby foreground and background sounds can be easily identified (Wrightson, 2000). Sounds intersect with each other and resonate in terms of frequency and rhythm. In this way, a sonic rhythmical balance is created, comparable to that of a natural soundscape (Krauser, 1993; in Wrightson, 2000: 11).

In a rhythmically harmonious way, foreground and background sounds generate a distinctive acoustic colouration and horizon. Acoustic colourations are 'caused by echoes and reverberations that occur as sound is absorbed and reflected from surfaces within the environment' (Wrightson, 2000: 11). They foster a sense of *place* via significant information that relates the listener to the physical nature of space and its scale. Moreover, as sounds emanate from the listener's own community, they 'express a community's identity' (ibid.). Singular acoustic colourations have an acoustic horizon of extended reach that helps listeners to situate themselves and gain a sense of space and direction, offering sonic distinctiveness and 'tonality' (Winkler, 2000: 29) to places.

The *sonic experience* of place-temporality, formed out of these unique and harmonious acoustic colourations, is principally characterised by *repetition* and *resonance*. *Repetition* is 'the feeling of reappearance of sounds occurrences perceived as identical' (Augoyard and Torgue, 2005: 97), and a key factor in the definition and perception of context. The typical repetitive sound units of a particular place make an environment recognisable, perhaps either as a familiar, peaceful, hostile or aggressive setting. *Sound resonance* manifests itself through the acoustic colourations of the soundscape which results from a system of standing, sinusoidal moving waves and resonance frequencies that continuously sound and stabilise in urban space. This resonance varies according to the degree of evenness of building facades, the contours and scale of the location, and is sustained by multiple reflections off the surfaces of the buildings, object or walls. In this way, sound repetition and resonance further significantly enhance the character and experience of temporality in urban spaces.

Fitzroy Square is characterised by a unique soundscape, constituted by a familiar group of background sounds typically heard throughout the day: for example the rhythmic shoe-tapping as people pass through the square, their murmuring as they group and meet or as they speak on mobile-phones, the slam of thick wooden-doors as people go in and out of office buildings, the braking of taxis and the banging of their doors as they drop and pick up people from the square. Other foreground sounds which announce the arrival of a new season or particular weather event at the square include those of autumn, like the shuddering of leaves in the central garden's trees and the wings of pigeons and other birds flying out in large groups; and those of spring and summer, like the music of evening concerts and the garden parties of residents. These sounds resonate and alternate in patterns of intensity, duration and repetition throughout the days, weeks and seasons, providing distinctive acoustic colourations and a unique sonic experience at Fitzroy Square.

The *rhythmicity* of place-temporality

Soundscape studies highlight the importance of concepts of rhythm, repetition and resonance in determining the sonic experience of place-temporality. Similarly, and expanding on this, it can be argued, that the rhythmic expression of temporal patterns of events and routines in urban place generally shapes the sensorial experience of place-temporality.

The temporal dynamics and patterning of social space by routines of everyday life has been acknowledged by a range of scholars. Torsten Hagerstrand (Lindqvist and Snickars, 1975) focused on the repetitive everyday life activities that meet, interweave and shape temporal location in time-space (Pred, 1977). Anne Buttimer spoke of event-routines that synchronise within a time-space framework, and referred to social space as a 'multilayered dynamic complex' constituted by time-space rhythms (1976: 287). David Seamon (1980) referred to social space as a temporal choreographed whole of body-ballets and time-space routines, as constituted by synchronised patterns of human gestures and everyday activities. And Zerubavel (1981) explored social life as repetitive, ordered and essentially rhythmic.

Figure 4.1   Leece Street, Liverpool
Source: Photograph by the author.

This social temporal rhythmicity of the everyday can be perceived in urban space. Lefebvre suggests that urban places are polyrhythmic fields of interaction, shaped by repetitive social practices and events that are overlain and harmonise in time and space (2004: 16). Lefebvre also reminds us of the relationship between rhythm and time, as in music for example. He introduces rhythm as constitutive of time and space, and the means through which time is both experienced and represented in the city. These social everyday rhythms can be not only heard but also visualised in urban space. They characterise places socially and shape and affect the sense of time.

Social everyday life in the city, however, is bound to nature and physical space, themselves temporal and rhythmic entities, and cannot be understood separately. Accordingly, an everyday sense of order and tempo is not only shaped by social rhythms but also conditioned by the rhythms of nature and architectural space. Conversely, places are not just social or architectural formations (Avarot, 2002) but temporal milieus, hubs of recurrent and synchronised stimuli of different kinds, everyday social routines, patterns of movement and other sensuous practices, circadian and seasonal cycles of nature, and visual and haptic patterns of physical space (Wunderlich, 2007; 2008). These rhythms can be sensed and tangibly engaged with in urban space. They are place-specific and unique in the ways in which they are temporally superimposed upon each other.

Figure 4.2   Piccadilly Circus, London

Source: Photograph by the author.

Places are symphonies of events. Specific, unique place-rhythms, as in music, offer urban places temporal structure, metrical order and pulse. Through this, they organise time, and set the perceived *tempo* of a place. Place-rhythms are also sensual and affective as they synchronise in time and space, shaping aesthetic rhythmical continuums and offering temporal distinctiveness. Accordingly, they shape the *timescape* of urban places and define the unique aesthetics of place-temporality. As a result of place-rhythms, everyday place-temporality tends to be characterised by a sense of pattern and rhythm, a sense of balance and resonance. Synchronised ensembles of distinct movements and social practices, natural and physical patterns define the everyday *timescapes* of urban places as shown in Figures 4.1, 4.2 and 4.3.

Place-temporality can thus be defined as a vivid and coherent sense of time that is rhythmically expressive, suggested by the orchestrated patterns of local place-rhythms, rhythms of everyday social life, nature and space specific to urban places. Place-temporality is a sensual and affective process, involving performance and representation, and distinctive place-rhythms characterise the sense of time in urban space, providing balance and resonance, continuity, and sensual and affective distinctiveness.

Figure 4.3   Fitzroy Square, Summer 2005
Source: Photograph by the author.

Fitzroy Square is an everyday place in London that is rhythmically rich and distinct. Firstly, it is a rhythmic architectural environment, with robustly patterned building facades and public space design. Secondly, it is a square where the rhythms of nature are vividly perceived. Daily and seasonal life cycles of nature express themselves profoundly in the central garden of the square. Thirdly, Fitzroy Square has a rich socio-cultural profile and is consequently a vibrant social place defined by multiple and distinctive social everyday rhythms. It is a place people choose to cross on their way to work and also a destination, a place to meet up and to spend time, socialise, rest and play. It is a residential and communal square, where everyday regular practices of residents and community activities take place. Furthermore, it is a multicultural place, where groups from different cultural and religious backgrounds (including Indian, African and Muslim groups) and other particular urban groups (such as hoodie-wearing gangs of youths), come together and repetitively express and represent their differences through unique dress codes and rituals. By contrast, it is also a workplace and venue for several well-established firms and institutions, and thus a place where daily work schedules and timetables are perceived and represented by the timed fluctuations of social activity and collective gatherings of white-collar workers. Lastly, Fitzroy Square is the venue for a varied cultural agenda of both private and public events, especially in spring and summer. During this time, and in contrast to many other squares in London, the square opens its gardens to the public. As a result of these regular seasonal events and garden openings, social everyday life is transformed and enhanced during half of the year. Fitzroy Square is thus a place where multiple place-rhythms superimpose, organise and induce senses of time, shaping place-temporality in distinctive ways.

Conclusion

This chapter is a brief introduction to the notion of *place-temporality* and its aesthetic and sensual significance have been examined in a particular urban space. The aesthetics of place-temporality is experiential and performative, and thereby comparable to the aesthetics of *the art form*, with praxis and performativity sitting at its core. Place-temporality is also an everyday aesthetics, made up of both flow and event experiences, and further characterised by slackness, inchoateness and drift, and as a sensual experience, is characterised by a vivid sense of time, a strong sense of flow and the experience of a distinct soundscape. A sense of flow involves a distorted sense of time, but also a sense of order, regularity and tempo, as is suggested by the patterned practices and events that one unconsciously engages with in everyday urban spaces that synchronise to form a particular tempo. The distinct soundscape of place-temporality possesses a rhythmic balance and qualities of repetition and resonance. Both the sense of flow and distinct sonic experience indicate that place-temporality is rhythmically expressive. This rhythmicity is induced by recurrent stimuli, social practices and events of nature and physical space, and these place-rhythms are unique, heard and seen as they orchestrate in time and space to shape the *timescape* of urban places. Sensual and affective, such place-rhythms offer order, regularity and structure to place-temporality and affect the apprehension of time in urban space.

PART II
Resisting Rhythms

Chapter 5

Urban Outreach and the Polyrhythmic City
Tom Hall

Introduction

This chapter draws on urban fieldwork research into the practice of pedestrian patrol. The empirical focus is on a small team of professional outreach workers operating in the city centre of Cardiff, Wales. The team works with some of the more vulnerable and troubling occupants of the city centre, including the rough-sleeping homeless. My interest is in the physical and moral geography of the city and in pedestrian patrol as a particular way in which to see and know urban space. Thinking about rhythm, I argue, is a useful way in which to approach and appreciate the work of outreach in the city. Outreach workers on pedestrian patrol are engaged in an elementary rhythmic practice: walking. As urban walkers they take their place alongside countless others on the city streets, but they are walkers with a particular purpose and their professional practice has a rhythm of its own. Navigating the city they must also navigate the urban rhythms with which their clients are (sometimes) so very out of step.

From the window

We can start with looking, and the look of the city.

In the opening lines to his classic study of city form, Kevin Lynch notes that '[l]ooking at cities can bring a special pleasure, however commonplace the sight may be' (1960: 1). Nor is this simply a matter of buildings and their spatial configuration. Moving elements, particularly people, are important too. Yet as onlookers, '(w)e are not simply observers of this spectacle, but are ourselves a part of it, on the stage with the other participants' (ibid: 2). Henri Lefebvre was similarly wary of seeing the city only as spectacle. Yet he suggests that a certain exteriority is necessary if one is to grasp the rhythms of urban movement: not an absolute withdrawal, but a balance of inside and out, just enough distance to gain perspective; '(a) balcony does the job admirably, in relation to the street ... [i]n the absence of which you could content yourself with a window' (2004: 27-28). Here he is, at the window of his Paris apartment on the rue Rambuteau, looking down:

> [t]he arrival of the shoppers, followed shortly by the tourists, in accordance, with exceptions (storms or advertising promotions), with a timetable that is almost

> always the same; the flows and conglomerations succeed one another: they get
> fatter or thinner but always agglomerate at the corners in order subsequently to
> clear a path, tangle and disentangle themselves amongst the cars. (ibid: 30)

Lefebvre's description of the city street, the surging passage of pedestrians amidst
the motorised traffic, '[f]rom right to left and back again ... passers-by ... buses
cutting across, other vehicles' (28) belongs to Paris but reaches well beyond. This
is the swarming anonymous life of the modern city, and the street. The narrow
constancy of the historic urban neighbourhood, 'a village life, or something like
it', overtaken and fragmented by 'dizzying swarms', rhythmic in aggregate but
made up of individual trajectories such that 'everyone has some route of his own
(from flat to school, the office, the factory) and does not know the rest very well'
(Lefebvre, 2003: 151).

And here is another window, in another European capital city: an office window
overlooking Central Square in Cardiff. Looking out and down, and talking, is Steve
Hyde, the manager of Cardiff County Council's City Centre Team (CCT):

> If you think about the city centre – and it is a busy space, Tom – there are some
> people there who are more or less invisible. You see them all the time but you
> might not notice them. In some ways it's like they don't belong. But they do.
> They do belong there. In fact they're there all the time, more than anyone really.
> It's a bit like they're ghosts: they're there, but you might not see them. Really
> though it's more like everyone else that's a ghost. Everyone else – ordinary
> people, going to work and shopping and going home again, and tourists, or
> whatever – they're like a blur really. It's like if you took a film and speeded
> it up, those are the people that would disappear. They're the ghosts. The ones
> you're left with, the ones who aren't moving and are there all the time, those are
> the people we work with. It's like there's two city centres really, two worlds.
> Only one of them is hidden away a bit. And there's not many of them. Cardiff is
> a big place, but what we're dealing with basically is a village, if you take away
> all those other people.

Not Paris, but the same urban types and tropes in play; the crowd mobile and
indifferent, individual pedestrians set on particular trajectories (work, shopping,
tourism) and attentive to little else. Only here in Cardiff, stood still amidst the
'ordinary' traffic and going nowhere are the city centre homeless. Steve does not
label this group as I have – as 'homeless'. Instead, looking down from the window
of the CCT offices, he invokes the potent urban imaginary of a world unseen but
under our feet all along, two realities socially distant but sharing space, the outcast
urban poor fading into the fabric of the city (for a literary example , see Ackroyd,
2001: 617).

In what follows I take these two visions of the city – rhythmic and mobile,
divided and unseeing – as points of orientation, and turn to look more closely and
at street level at the work of social care that Steve and his team undertake in a city

in which the look of things, the spectacle of the city, has come to the fore in recent years in the context of urban regeneration.

On the street

Cardiff was one of the great maritime ports of Victorian Britain, although never a major industrial conurbation; today it serves as an area hub for public and service sector employment in particular, an 'enclave of relative prosperity in one of the poorest regions of the UK' (Thomas, 2003: 12). Cardiff's commercial, civic and administrative core occupies not much more than a square mile in the heart of the city; it is a space described by railway lines and a river – the London to Swansea railway line (flanking Central Square), the river Taff and the Valley Line service to Pontypridd. This is the patch that the CCT routinely works, which is to say that their remit is spatial as well as social. The team works with and supports vulnerable people in the city centre. Potential clients are those whose behaviour, addictions, health and housing needs, financial, family and other circumstances or difficulties have brought them into some sort of a relationship – a spatial relationship – with the centre of the city; people who have come unstuck in the middle of things. This includes street drinkers, beggars, prostitutes, drug-users, and the homeless, particularly rough sleepers. As Steve suggests, there are not so many of these, no more than might be found in other UK cities; but they are there (for those who can see), lost amidst the rush, out of place and going public with their troubles. The CCT works with these and other clients as a multi-disciplinary team, bringing together variously specialised social workers , a number of support and advisory staff and project workers, an NHS-funded nurse, and three outreach workers. It is these outreach workers and their space of operation that I focus on. This focus is not arbitrary, for outreach is a fundamental component of the wider work of the CCT. It is through the work of outreach that the team makes and maintains contact with a core client group some of whom would otherwise struggle to access mainstream social and health services. When interested parties ask Steve about the work that his team does, the first thing he offers is an invitation to accompany the outreach workers on duty. Not because that is all there is to the CCT, but because that is where the work begins, with pedestrian encounters, out of doors, with difficult, needy people. Like this:

> Early morning, late October – just gone seven o'clock, and still dark. I am with Nicy, a member of the CCT, and we are doing outreach together: we are on outreach. Specifically, we are taking a tour of central Cardiff toting flasks of tea and coffee, a couple of dozen bacon rolls wrapped in foil and a tupperware box of hard-boiled eggs. This is the Breakfast Run, an early-morning outreach patrol distributing hot drinks and food to rough sleepers; it is something the CCT do every weekday morning.

Having crossed the river we are now taking a walk – a nosey – though Central Square and the bus terminus. At this hour we're still looking for sleepers, but the first person we find is already up and about: Vincent. Most nights Vincent sleeps in the EOS, Cardiff County Council's 'Emergency Overnight Stay' accommodation, a single room in a larger hostel building, converted nightly into bed-space for six on adjacent rubber mattresses laid out on the floor. EOS operates on a first-come first-served basis, and Vincent must sign himself in every night. Last night he was too drunk to make it, and slept out in the bus terminus instead. Not for the first time. This morning he is not at his best, in fact he looks a wreck: derelict, slurred, slow with the cold, confused. He is sat on the pavement next to a puddle of loose change he has spilled out of his pockets – copper (lots), silver (lots), a few pound coins. He is sorting the money into piles it seems but the process is very slow and clumsy; his movements thick, and his hands twitching and juddering. (Hung-over is not really the word for Vincent's condition; he is very ill with drink and 'rattling' as he would call it.) He smells powerfully of sweat and alcohol and urine.

Nicy tries to get Vincent to come round the corner to where the drinks and food are stashed, but it is hard to tell if he is making any sense of what she says, hard to tell if he will be able to walk either.

'Do you need to get a drink first, Vincent?' Nicy asks, indicating the heap of coins.

'They won't serve me,' he whispers. The (licensed) convenience store on the square won't let him through the door, the state he's in.

Nicy is brisk and positive. 'Oh. Well, would you like a cup of coffee then?'

A long pause and Vincent indicates yes, and starts, painfully slowly, to gather his coins: one coin at a time.

Nicy has crouched down to his level, on the balls of her feet, keeping her balance with one hand on the floor. This sorting of coins will take forever. Vincent mutters that he was robbed last night when asleep, he lost £60 odd.

'No,' says Nicy. 'Oh, Vincent. After all that bother.' (And there was a lot of bother, and Vincent was the cause of the best part of it – raging and drunk yesterday and refused entry to the benefits office and then to the CCT offices in his pursuit of a giro he thought he might be due and which Nicy helped him secure, after several hours of drama.)

Nicy asks Vincent, with deliberate kindness, if she can help him get what remains of his money together, not wanting to touch without his permission. He nods assent, and she scoops his money off the pavement and tells him to hold out his hand. He does but is rattling badly.

'Put your hand on mine, Vincent,' says Nicy, holding out her free hand, palm up.

Vincent does so, his hand cupped too, resting on hers and held lightly in place as Nicy pours the coins on. Vincent pockets his change, with a little help. He struggles to his feet, and follows us as we walk round the corner, where another regular is already waiting. We dole out tea and coffee and rolls (Vincent says he can't eat anything but Nicy makes him take one, instructing him to keep it for later). We leave these two and head on to Bute Park, Nicy rubbing her hands clean with antiseptic gel.

This is outreach at its most raw and unelaborated. Nicy reaches out, makes contact, and gives something, on the street. Of course, work will follow from this, or could do. Back at the office later that day Nicy will make some calls, add notes to a file, speak to Vincent's social worker; she may look for him again later that day and walk him over to the EOS if he is amenable. Vincent's name will come up for discussion at the weekly team meeting. A good share of the wider workload of the CCT entails the more mundane and mainstream business of public sector welfare provision – calls, case-notes, needs assessment, computer entry, client visits, referrals, team meetings. Even so, at the back of it all, and crucially for the CCT, there are outreach workers, like Nicy, walking through the bus station in the early morning; a daily labour of looking in the city, with an eye for those hidden away, overlooked and out of sync with the predominant urban rhythms of use and space.

Across a border

Where does this work of outreach take place? In the city centre and on the streets; this is work that cannot be done from a distance, at a window. As Lefebvre asserts, hypotheses suggested at the window are to be explored – confirmed or invalidated – on the street (2004: 33). Steve's outreach workers are out there to step across the boundary he describes: 'two city centres really, two worlds ... one of them is hidden away a bit'. This is what makes outreach work special to those that do it. Leaving the office, outreach workers walk out on a world of bureaucratic administration and enter into working relations that mark what they do as different. Nicy meets with Vincent on his own terms and territory, works on her feet, does not keep office hours (more of which below); the work is something others might recoil from, a physically and morally unsettling, sometimes dangerous, undertaking (see Rowe, 1999); it requires a particular commitment and something of that same eye for the urban underside that Steve exhibits. Outreach workers must (learn to) see a different city. They have to home in on just those people that others overlook or avoid, and learn to pick up on a variety of clues and cues distributed across the cityscape: carefully folded cardboard sheets tucked away in a known corner, a slat of wood prised off a boarded window, scorched foil wrapping and discarded chocolate bars round the back of the train station. When I started to register the same little signs myself, after some months spent accompanying outreach patrols in central Cardiff, I was pleased to be told 'Oh you've got outreach worker's eyes now, Tom'. There is a measure of romance to all this, something of the same thrill that informs social exploration as a journalistic and literary endeavour – but which outreach workers must hold in check. Imaginative sympathy and identification must be balanced with a professional approach and practical objectives. As Richard Sennett has it, writing of social work more generally, strangeness and connection have to be weighed against each other in an attempt to respectfully bridge a 'boundary of inequality' (2004: 20). Yet this is a boundary that outreach workers in Cardiff must do more than bridge. The provision of immediate services and a little human contact early in the morning is one thing, but the real ambition is always to bring people like Vincent back across and on side – into accommodation, out of trouble, off drugs, away from all that. Off the streets. In this way, the line that outreach workers patrol is not just the contour but also the measure of the work that they do.

This is the figurative geography of outreach; a socio-moral perimeter that runs, contrarily, through the middle of the *physical* city. Outreach has its own landscape, an environment which outreach workers can properly discern and make 'tell' (see Ingold, 2000: 190). It also has a rhythm.

On the beat

I have described the CCT outreach work as a labour of looking in the city – CCT outreach workers spend their working hours scouting for need in the urban centre. As such, I argue that the work of outreach is fundamentally rhythmic, drawing attention to three attributes. The first is the pedestrian character of the work. Outreach workers are walkers, nothing if not street-level bureaucrats (see Lipsky, 1980), and walking is an essentially rhythmic practice – fundamentally so, as an elementary bodily motion realised through a regular, repeated, sequential transfer of weight. The rhythm of walking could be considered universal, wherever it might take place or deliver us; and also, in effect, timeless: a vital and enduring component in the bodily history of bipedal evolution and human anatomy (Solnit 2001: 3). Yet we walk with, and to, different ends, as I will discuss. In describing Nicy's encounter with Vincent in the bus station, I have emphasised the caring rather than the investigative and itinerant character of the work – my principal aim was to take the reader down from the window to the pavement, and to convey something of the immediacy of the work of outreach (and signal the boundary of want and difficulty across which Nicy and her colleagues practice). But it would be a mistake, extrapolating from this one vignette, to construe outreach work as a series of such (stationary) encounters, with Nicy and her colleagues setting down to work in one location and then another. To portray outreach work as a sequence of 'stops', in this way, would be to misapprehend its essence as a mobile practice, a roving administration. Operating at street level, on foot, outreach workers join the rhythmic eddies of people and traffic that Lefebvre depicts. They do so in a particular way, threading their own passage through the city centre. However, though this is a passage aligned to a professional purpose, it is not one that could be described as typically purposive. The aim is not so much to efficiently transport the walker between sites as to roam the cityscape in an open and exploratory mode at a pace aligned to general enquiries. Outreach workers dally; they meander and snoop, open to chance encounter and diversion; they seldom take the shortest route between two points; they are, perhaps, latter-day flâneurs – hardly dilettantes, because at work, but, even so, drifting unhurriedly through the city the better to know it, taking in the look, feel and tempo of the streets. Outreach, then, is walking as a discursive rather than purposive practice. The distinction between these two (and other) modes of walking in the city is made by Wunderlich (2008) in a wider discussion of walking, rhythmicity and urban space.

Repetition is the second rhythmic attribute, for outreach is a repetitive practice and, as Lefebvre notes, there is '[n]o rhythm without repetition in time and space, without *reprises*, without returns' (2004: 6). To accompany the CCT outreach staff over any period of time is to be apprenticed to a soon familiar schedule of repeated activity and routes traced through city space. Three sweeps of the centre of Cardiff constitute the work of outreach on any given day. The first of these is the Breakfast Run, the early morning patrol that brings Nicy and Vincent together. Much later on, about 5.30 pm, a second patrol sets out to spend two or three hours moving through the urban afternoon on the look out for much the same combination of need and trouble that Vincent represents. Later still there is a third and final sweep targeting the city's red light areas, with outreach workers walking a beat alongside the city's street sex-workers – making contacts, building trust, distributing condoms and hot coffee and health advice – finishing up at about half past ten at night. Early the next morning the sequence starts again. The regular timetable is matched by regular patterns of movement through space. Early morning outreach follows much the same route each day, tracing a path along a known itinerary of 'homeless' sites: the bus station, Bute Park, the museum steps, the NCP car park, the subway. Afternoon outreach takes a different shape, similarly repeated, with workers checking off a known sequence of sites and settings as they move through the city. Cardiff's red light areas constitute an established geography and here too the spatial outline of outreach on any given night will be much the same as on any other. The work repeats itself, in time and space, as a sequence of scheduled patrols operating along customary routes. If outreach work is in this way repetitive, this is not to say that it lacks interest and excitement. Each working day presents new challenges, and the client group is varied and in some respects unpredictable. Seldom a day goes by without some flurry of excitement or crisis to which the team must quickly respond, dropping settled plans and dispatching workers to wherever the trouble might be. Added to which, outreach itself is exploratory and inquisitive. If outreach workers patrol the city at more or less set times and along familiar paths, they also digress, straying off route to act on particular information received or just to 'take a little nosey over here'. Even so, the work is nothing if not recurrent: same times, same places. The baseline itinerary is always there, and to be relied upon. Whatever happens in the course of a late afternoon outreach patrol, and wherever this takes the workers, the chances are that they will end up, as usual, round the back of Marks and Spencer's mingling with the crowd that always gathers to meet the eight o'clock soup run at which church volunteers distribute drinks and food and sometimes clothes to all-comers.

Lefebvre makes a distinction between linear and cyclical repetition, the former indicating social practice 'the monotony of actions ... imposed structures', the tick-tock of the clock, and the latter originating 'in the cosmic, in nature: days, nights, seasons' (2004: 8). Closely aligned, although not identical, is Tim Ingold's distinction between metronomic and rhythmic repetition. 'A metronome, like a clock, inscribes an artificial division into equal segments upon an otherwise undifferentiated movement; rhythm, by contrast, is intrinsic to the movement itself' (2000: 197). I suggest that the repeated movements of outreach workers through Cardiff are rhythmic rather than metronomic. The schedule of patrol I have described above does not result from the imposition of a timetable upon the activity of outreach, a structure to which the work is then fitted; it arises from the work itself, which takes (and moves through) place(s) at times in accordance with the patterns of use and need practised and experienced by its core client group. Thus early morning outreach meets a moment in the circadian cycle when the city centre homeless are at their most physically and emotionally vulnerable (certainly so in the winter months), when chronic drinkers are at their most sober and suggestible, when human bodies, ready to break the night's fast, are receptive to the kindly but sided advances of outreach. A quiet word about taking up a place on a rehab programme generally finds better purchase at first light, over a hot cup of tea, in the calm of a dark December morning, than it does at midday. Late afternoon outreach has its own logic also. As office hours come to an end the ghosts and stragglers that Steve observes from his window '[t]he ones you're left with, the ones who aren't moving and are there all the time' are easiest to spot, remaindered as they are by the streams of commuters and shoppers exiting the city centre. Again this is a 'low' moment, which outreach workers look to exploit: with night coming on and the city winding down even the more difficult and 'chaotic' (see Masters, 2006: 2-4) among the CCT client group may become amenable to advice and assistance. Finally, the third sweep sees outreach teams appear on cue alongside the first of Cardiff's street prostitutes. This aligning of outreach patrols to the lived tempo and circumstances of the client group gives outreach its rhythm and provides a safety net for those who sometimes struggle to keep to the metronomic schedules of mainstream welfare provision: 11.20 a.m. appointment at the Housing Advice Unit; GP's surgery at 14.45 p.m.; Benefits Agency interview, 15.35 p.m.

Cardiff, like all cities, is polyrhythmic; the rhythms of interaction between CCT workers and their client pulse beside myriad concurrent others, each generated by particular priorities and pressures, by particular groups and intersections of interest. Nor do these rhythms operate independently. This leads me to the third rhythmic attribute of outreach work in Cardiff. I suggest that the CCT and its clients share dominated rhythms within the polyrhythmic city, and also that the root ambivalence of outreach work, which seeks always to care but also to manage and manoeuvre, implicates its workers as informal agents of social control. Outreach is a rather unsung and lowly occupation – pedestrian, not terrifically well paid; and its client group has little standing in the city. As such, outreach workers have to align and subordinate their practice to the wider tempo of the city. Having described the crepuscular pattern of outreach as rhythmic rather than metronomic, arising from the work itself, I now note that the CCT can find itself under considerable external pressure to keep to the shadows. The special pleasure that looking at cities can bring (Lynch, 1960; see above) is soured for some if the scene is conspicuously peopled by derelicts and druggies, and in Cardiff, particularly, the longstanding 'professional and political consensus about the importance of creating a city centre more fitting to the city's image as the capital of Wales' (Thomas, 2003: 18) makes the look of the city particularly important.

In this context, CCT outreach is a tolerated (and valued) practice only so long as it does not intrude on and disrupt the public and commercial rhythms of the city centre. Early morning outreach is fine, so long as Nicy and Vincent are both gone before the bus station gets busy, so long as no one is handing out hot coffee on the museum steps once the museum has actually opened. Outreach is best practised out of hours, when the needs to which it caters are less likely to be 'seen'. Indeed Steve and his team are continually flexing the contours of CCT outreach to align these to the assorted requests and demands that come in to the CCT office from any number of constituencies with a stake in the city centre, to which they usually defer. Not only is the CCT obliged to align its practice to other and dominant urban interests, doing the best it can to cater to a marginal client group without allowing this work to disturb the pulse of the city, it also operates as a soft agent of social control. Steve and his team straddle an ambivalent agenda tasked to befriend and support but also to supervise a needy and sometimes difficult client group, and as 'supervisors' of need and difficulty they are sometimes called on to help police the city. Thus the CCT office can field a variety of calls throughout the day, in between the usual outreach sessions, asking for assistance in dealing with a 'problem' client, and will routinely dispatch an outreach worker to try and 'sort' whatever difficulty has arisen. In such cases, the call to the CCT is commonly prompted as much by the public disruption caused – by a man urinating outside the cinema, or a drinker haranguing passers-by on the car-park stairwell – as by any particular concern for the individual. What is required is that the problem be taken care of, and removed. In this way, rhythms of city space and use are, again, at issue. Any number of young women walk around the city centre late at night in their bare feet, high-heels dangling from straps, and drunk, but if a woman is weaving her way through the central train station ticket hall barefoot and incoherent in the middle of the day then that is something else, and the chances are that the transport police will give the CCT a ring to see if this is 'one of theirs' and if something can't be done. Minor instances of urban arrhythmia are what set the CCT phones ringing. The CCT first encounters a number of its eventual clients this way: a call from a concerned party – police, proprietor, passer-by – about a person seemingly, conspicuously, out of sync with the city.

Out of step

Tasked to keep the ragged fringe of the urban (social) fabric from unravelling altogether, the CCT carry out a job of routine maintenance in the city centre akin to the workaday 'hidden' labours which see the material city maintained (Thrift, 2005). They are a call-out and/or clean-up crew, combining elements of emergency response with routine maintenance and upkeep; and like all maintenance operations, this work goes on and on. Outreach patrols cover the same ground again and again, literally and figuratively, walking the same beat and patching up the same (sort of) problems. The poor are always with us, after all, and with the city – which is not to say that nothing matters much, or changes. As Zygmunt Bauman notes, what it means to be poor can shift considerably, depending as it does on the kind of 'us' the poor are 'with' (1998: 1). It also depends on the city they are within. And Cardiff is changing, which is to recognise a further, final and deeper rhythm: the pulse of economic transformation, of urban regeneration.

Recent years have seen the city of Cardiff transformed – reborn even (see Ungersma, 2005) – in the context of the new urban economic priorities of service-sector work, shopping, leisure, tourism (Bristow and Morgan, 2006; Punter, 2006). The regeneration continues apace, following an essentially boosterist strategy in which the attractions of a redeveloped city centre prime the pump for reciprocities of commercial investment and retail spend. This effort entails not only a redeployment of people, energies, and finance, but a redeployment of land (Byrne, 2001: 29), its use, design and look. Thus the view from Steve's window has transformed in the time that I have involved myself with the work of the CCT. The 'rather tired looking and overcrowded central bus station' (Thomas, 2003: 40) is still there, thronging with rhythmic crowds, but is now ringed by demolition sites and plywood hoardings promising new, luxury city-centre apartments. A few hundred yards away, across the rooftops, the nautical masts of the Millennium Stadium profile the city as an 'international visitor destination' (Punter, 2005: 100). And behind the CCT office, beyond the multi-storey car park (also scheduled for demolition; its back stairwell only too well known to Nicy and her colleagues and their clients over the years), is a 967, 500 square foot construction site in the heart of the city, replete with cranes and bulldozers, stacked prefab units, hard-hats and towering frameworks of I-beam and rebar steel, soon to be a new shopping centre offering 'attractive and safe public spaces for people to enjoy, alongside a unique and exciting contemporary retail space which caters to the needs of the 21st century shopper' (see http:www.stdavids2.com). These developments have their consequences for the CCT and its client group and the work of outreach in the city centre, ratcheting up existing tensions and disrupting the delicate, negotiated, eurhythmic accommodations that Nicy and her colleagues have made in fitting their labour of looking and care (and control) to the public and commercial rhythms of the city.

Cardiff's regeneration is rapacious and image-conscious, and if Vincent still has his place in the early morning quiet of the as yet un-redeveloped bus terminus then this may not be for long. Elsewhere, across the city centre, he is now 'seen' and stands out – in the regenerated public space of precincts and café quarters; an expanding night-time economy has similarly illuminated and exposed the shadows in which Vincent might once have escaped notice – that second, 'hidden' world that Steve portrays. Less and less is out of the way in the middle of Cardiff today, and it is getting harder and harder for outreach workers and their clients to find time and space for one another in the centre without rubbing others – the rest of the city – up the wrong way.

Conclusion

In closing, I return to Kevin Lynch and the look of the city. Noting the special pleasure that comes with looking at cities, Lynch draws attention not just to the spatial configuration of buildings but to the mobile life of cities – the traffic and crowds, the people. He adds that we are more than observers of the urban show. Our own movement through the cityscape implicates us: '[w]e are not simply observers of this spectacle, but are ourselves a part of it, on the stage with the other participants' (1960: 2). Henri Lefebvre, who certainly took a special pleasure in looking at cities, suggests that this mobile life might be understood as a polyrhythmic whole. Looking out at the city from the office window of Cardiff County Council's City Centre Team, the CCT manager, Steve, reminds us that a shared stage is not without its divisions, its winners and losers. Some of those on their feet in the city are going nowhere. Taking these as points of orientation, I have used this chapter to look at the intersection of two groups of people in the city of Cardiff, outreach workers and their homeless and otherwise difficult and vulnerable clients, attending to their mobility across and shared occupancy of the city stage, and in particular to the ways in which these can be seen and appreciated as rhythmic. I have described CCT outreach in Cardiff as a marginal activity – lowly, unregarded, invisible – which nonetheless moves through the middle of the city. As such, it has to accommodate itself to the dominant public and commercial tempo of the city. This work of rhythmic accommodation is continually (re)negotiated, but becomes especially acute, I have suggested, in the context of urban regeneration aiming to enhance the look and spectacle of the city. In Cardiff today the very stage set is moving, not just the people; and the new spaces created look to be brighter and more welcoming, but not simply so, and not for all.

Chapter 6

Fascinatin' Rhythm(s): Polyrhythmia and the Syncopated Echoes of the Everyday

Deirdre Conlon

Introduction

Henri Lefebvre's fascination with rhythms was evident through much of his work, but a compilation of insights on the significance of rhythms for understanding the interlocking temporal and spatial character of contemporary social life only appeared as the finale of his written work. *Éléments de rhythmanalyse: Introduction á la connaissance de rhythmes* (1992) was published in French a year after Lefebvre's death with a translation to English first published in 2004. For those who – like me – are captivated by and engaged with Lefebvre's ideas this book presents an invitation to chart intersections between *Rhythmanalysis* and some of his writings preceding this work, to elucidate a productive coupling of temporal *and* spatial elements that comprise lived space and social life in the modern world and thus to develop a more comprehensive account of the importance of Lefebvre's work for contemporary social and cultural geography.

In this chapter, I address the following particular points: first, I retrace some of the overlapping terrain between Lefebvre's critical interest in everyday life and the analysis of rhythms; then I highlight how a focus on the polyrhythmia of everyday spaces informs an understanding of Lefebvre's perspective on historical time, most notably, the critique of a chronological sequence and so-called progressive rendering of history. The everyday spaces I discuss are the 'once' slow-paced rural towns of contemporary Ireland. Ireland has recently undergone a momentous economic and socio-cultural makeover in association with the rhythms of global economic processes, demographic flows, and attendant shifts in consumption, commodification and lifestyle. By homing in on the rhythms of change, repetition and difference that are felt and expressed, I suggest we can gain an appreciation for the multi-layered and interlocking temporalities that make this and other social spaces hum amidst the polyrhythmic chorus of the everyday.

Tracing the continuum: Rhythmanalysis and the everyday

Edensor and Holloway (2008), Evans and Jones (2008), Elden (2004a) and Mels (2004) have recently called attention to the diverse array of temporal metrics and pulsations – or rhythms – we encounter routinely and variably throughout our lives. This polyphonic array includes physical, linear, social, cyclical and biological rhythms that 'may clash or harmonize, producing reliable moments of regularity or less consistent variance' (Edensor and Holloway, 2008: 484) and confirms, as Lefebvre suggests, that 'what we live are rhythms' (1991: 206). Rhythms and everyday life are thus invariably intertwined, and within Lefebvre's oeuvre '*Elements of Rhythmanalysis* explicitly continues the project of the *Critique of Everyday Life*' (Elden, 2004b: 194).

The issues taken up in his *Critique of Everyday Life* 'assert Lefebvre's Marxist credentials' (Merrifield, 2006: 9) while simultaneously affirming his concern with broadening our apprehension of capitalist relations of production and reproduction in light of their integument across all creases of modern social life. To this end, Lefebvre elaborates on the ever-expanding disengagement and alienation of social practices from the processes and structures integral to their making, and couples this with a detailed investigation of the intensified commodification of everyday life. More concretely, for Lefebvre, the workplace is not the only site of exploitation and alienation in the modern world; instead, the sites and practices of the everyday, including family life, private domestic spaces, leisure activities and vacations, as well as travel between these social spaces, have also been subsumed within the folds of exchange value. With this, propensities to calculate and measure, as well as to exploit, everyday life have accelerated. As production and consumption have enveloped more of the everyday, the linear metre these processes obtain comes to dominate. Accordingly, Lefebvre's critique of everyday life resounds with the analysis of rhythms. The distinct yet inter-related practices of production and consumption are materialized as a sequence of passive, seemingly banal day-to-day repetitions characterized as *métro-bulot-dodo* or 'commuting, working, sleeping' (Moran, 2004: 54). Consequently, the rhythms of capital now dominate everyday life and work not only to alienate individuals from practices of production but also from themselves. As Lefebvre suggests, such 'repetitive gestures tend to mask and to crush the cycles […], nights and days, seasons and harvests, activities and rest, hunger and satisfaction, desire and its fulfillment, life and death' (Lefebvre, 1987: 10). The critique of everyday life thus foreshadows the project on rhythmanalysis and demonstrates Lefebvre's longstanding concern with the measures that mark the everyday.

Some of the scholarship on the everyday tends to reinforce dichotomies by focusing either on the alienating facets of modern daily life or by celebrating resistance to this banality and disaffection. This has led to charges that studying the everyday has become a wide-eyed fetish among academics (Felski, 2002; Colebrook, 2002). Gardiner (2004) responds to such critiques by suggesting that a close reading of Lefebvre makes clear that his rendering involved agitating these simple dualisms through a firm and materially grounded commitment to apprehending the complexities of everyday life. In *Rhythmanalysis*, this view is made more explicit as Lefebvre identifies and classifies a host of intersecting rhythms, including the polyrhythmic, eurhythmic, isorhythmic and even arrhythmic measures as well as secret, public, internal and external beats (2004: 16-18) that comprise the symphonic everyday. Several of the essays in *Elements of Rhythmanalysis*, perhaps most notably 'The Rhythmanalyst: A Previsionary Portrait' and 'Seen from the Window', also provide methodological guidance for examining the multiple beats that mark everyday social action, as I later discuss. Thus, as a continuation of Lefebvre's critique of modern life and as a culminating point in his life's work, the analysis of rhythms integrates the temporal with the spatial and moves beyond dualities. In addition, the analysis of rhythms suggests a mode of praxis for investigating the overlapping influence of these multiple measures in social space.

All this said, it would be remiss if I did not acknowledge that Lefebvre's examination of everyday rhythms strayed off-kilter on occasion. Despite the vibrant multiplicity inherent in their classification, his analyses were sometimes ensconced in bifurcations of linear or institutional rhythms vis-à-vis cyclical and/ or resistant beats. Edensor and Holloway (2008) pick up on this issue, and in their investigation of the multiple rhythms that embody the temporal spaces of a coach tour in Ireland, they push for more sustained attention to the ongoing materialization and fluid organization of the multiple rhythms we encounter daily. Recognizing Lefebvre's sensitivity to 'the polyrhythmic constitution of spaces, often including the non-human, technological and material' (2008: 486), they develop an analysis of how 'this co-evolvement is assembled and how rhythms produce an apprehension of unfolding space and time' (ibid.). In this chapter, it is the assemblage and dynamic interweaving of the present and the past within the rhythms of everyday life that I examine.

Some notes on syncopated beats and the residues of the everyday

In a section of *Rhythmanalysis* specifically devoted to 'music and rhythms', Lefebvre notes 'music [...] offers to thought a prodigiously rich and complex field' (2004: 55); I concur and want to extend the musical overtones that rhythmanalysis invites by employing the concept of syncopation as a way to figuratively grasp the dynamic co-existence and influence of the multiple beats that shape social space. To elucidate, all rhythms are comprised of a downbeat and an upbeat; two inseparable parts working to mark time and give momentum to a musical piece. More often than not, in musical spheres, the downbeat is given greater emphasis than the upbeat, but with a syncopated rhythm it is the upbeat, or the offbeat, that carries weight and moves the melody along. What's important here is literally a matter of emphasis; with regular and syncopated rhythms, the downbeat or the offbeat is alternately emphasized while both facets are always necessary to the overall shape and thrust of a piece of music. By extending this thinking to Lefebvre's concept of polyrhythmia in everyday life, I suggest we can productively circumvent the previously noted tendency to reproduce the analysis of rhythms in a dualistic way. To this end, this chapter details an ethnographic encounter with some of the dominant beats that mark contemporary Ireland including the sights and sounds of fast-paced development and rural cosmopolitanism. Then, homing in on syncopated beats, I examine how these dominant rhythms coincide with the slower paced, offbeat accounts given by asylum seekers whose everyday lives are frequently inaudible amidst the downbeat din of the Celtic Tiger.

While Lefebvre makes only one specific reference to a syncopated rhythm, pairing it with a continuous beat in a discussion of regular and irregular beats (see 2004: 69), he does allude to a parallel idea in noting that 'rhythms differ from one another in their amplitude, [and] in the energies they ferry and deploy' (1991: 206). By picking up on these ideas and calling attention to the downbeat and the offbeat, to the regular and syncopated rhythms that are always present and that concur in time and space, I argue that a more nuanced rendering and deeper understanding of the intersections between rhythms and everyday life is made possible.

Rhythms are also intertwined with Lefebvre's interest in recurrence and difference. These ideas inform his critical perspective on relationships between the temporality of history and the contemporary everyday. As previously noted, the linear rhythms of modernity can drown out other measures. But beneath the vale of abundance and innovation that modernity (re)produces through day-to-day routines, Lefebvre saw the 'everyday as a residuum [where] our culture drags in its wake a great disparate patchwork which has nothing "modern" about it' (Moran, 2004: 55). Consequently, if everyday life is crisscrossed with a multitude of rhythms, the singularity with which we typically measure time is thrown into question. As we attend to the multiplicity of rhythms that mark everyday life we confront patterns and practices that are more closely associated with bygone periods and this, in turn, pushes us to rethink notions of history and the chronological measurement of time. Moran (2004) puts the matter more succinctly: 'the traces of the past which are contained within the everyday undermine a continuist, processual notion of history' (2004: 56). Lefebvre asks, '(W)hat does the proximity between a certain archaism attached to history and the exhibited supermodernity whisper?' (2004: 34). Unravelling this question Lefebvre touches on the fluid interplay between old and new, between recurrences and innovations, that inhere in everyday routines. I explore these issues within the context of recent changes and the rhythmic remainders from times past that are evident in contemporary Irish society.

Setting the scene: rhythms of change in Irish society

Until recently, Irish society was commonly rather statically construed as one of the few places within Europe that stood apart from, or at the periphery of, the waves of market-driven development, connectivity and technological expansion that characterize globalized modernity. Associated with the slow pace of economic and political development and the uneven heave of social and cultural shifts through much of the twentieth century, emigration was an enduring feature and routine practice of everyday Irish life. Although there was a brief period in the early 1970s when this practice was reversed, with inward migration – primarily by returning Irish emigrants – exceeding the rate of emigration (see Gray 2004; Walter 2002), this trend was short-lived; thus, it is fair to say that for an extended period lamentations on emigration were the drone of the day, the year, the decade and chronologies beyond that too. More recently however, this scene and the long-standing daily rhythms embodied therein appear to have changed dramatically. Beginning in the mid-1990s and for roughly a decade, the boom and bliss of the so-called 'Celtic Tiger' became the dominant pulse in Irish society. Needless to say, the exhilaration that reverberated throughout Ireland has been severely tempered in association with recent financial crises and economic slowdowns. Nonetheless, from the mid-1990s through the early years of the twenty-first century, the nation's economic growth rate swelled, unemployment rates plummeted, wealth increased and consumption-driven, cosmopolitan lifestyles became the order of the day (see Bartley and Kitchin, 2007; Jones, 2003; O'Connell, 2001).

Coupled with these transformations, during the mid-1990s immigration to Ireland replaced the long entrenched history of emigration; net migration increased from 8,000 in 1996 to 19,200 in 1997 with further dramatic increases to 71,800 in 2006 (Central Statistics Office, 2003; 2007). Today, foreign-born residents constitute 10% of Ireland's population (Central Statistics Office, 2007); this population comprises returnees – former emigrants who have returned home – labor migrants from the new European Union member-states and individuals who are seeking or have been granted refugee status. Among these groups, I contend that for asylum seekers the rhythms of everyday life are entangled with the din of Ireland's recent encounter with global modernity in quite revealing ways. For them, the roars and rhythms of Ireland's 'Celtic Tiger' jaunt are rather muffled and everyday life exudes a different beat, one that seems to be more in tune with the past. By juxtaposing the routine day-to-day experiences of asylum seekers alongside some of the rhythms of change that are tangible just about everywhere in contemporary Ireland, we can garner a sense of both the upbeat as well as the offbeats that combine and resound the past within the polyrhythmic chorus of the everyday.

Rhythmanalysing: ethnographic encounters and interviews

I now recount some the things I discerned in the course of a study of women asylum seekers who had arrived in Ireland amidst the exuberance and afterglow of the Celtic Tiger years (see Conlon, 2007). In accordance with the Irish government's policy of housing dispersal for asylees, these women were living in an accommodation centre in the west of Ireland. As a result, my work with these women necessitated traveling through and to places I had not visited for many years. Consequently, the rhythmanalysis I speak of here took place within the context of my journey to meet and interview a number of women whose applications for asylum were pending review.

From a methodological standpoint a question remains: 'what do you think [and do] when you speak of rhythms?' (Lefebvre, 2004: 16). Lefebvre's response indicates that rhythmanalysis requires some of the tools of psychoanalysis, namely listening but without being passive, and some of the tools of science and statistics, specifically becoming attuned to measures of frequency, duration and the like, in combination with the acuity of an artist or a poet. Thus, the rhythmanalyst pursues 'an interdisciplinary approach' (22) that brings 'together very diverse practices and very different types of knowledge' (16). Throughout this process the rhythmanalyst takes his or her own body 'equally attentive eyes and ears, a head and a memory and a heart' (36) as a point of reference. To this end, the methodology I draw on is an ethnography of sorts; I recount some of the observations, sensations and reflections recorded in one midland town where I stopped en route to an accommodation centre in the west of Ireland. The effect I hope to produce is 'a scene that listens to itself' (36) and by combining this with asylum seekers' narrative accounts of everyday life, I hope to detail the 'bundle of rhythms, different but in tune' (20) that exists in the wake of the 'Celtic Tiger' boon.

Traveling west from Dublin on a weekend morning, it was easy to register the changed spaces that have materialized in association with Ireland's economic 'miracle'. En route to my destination I stopped in one midland town, which today is flanked for miles around with ex-urban, commuter-zone ribbon development along with the gray depots and warehouses of suburban sprawl. At the town's centre are a number of fixtures – a market square, shops, and the requisite pub – each helping to frame a social space with an ambience that is distinctly more urbanized and cosmopolitan than I recall from previous journeys. The market square, where not so long ago cattle were traded, has become a parking lot. With this, the rhythms embodied by cars and drivers are of speed and haste, periodic pauses and lurches into parking spots. Honking horns communicate intent, irritation and 'so long' as busy shoppers stop and shop, then rev and go. Like Lefebvre's observations from a window in Paris, the motions and meter, while synchronized, are continuously sporadic. Along with the cars is the pedestrian flow. Outside pubs, people – mostly men, both young and old – stand or lean while they smoke and talk; at café windows, couples pause to review menus and ogle at pastries. Store doors open and close as customers go and come, while waves of 'hello' and 'bye-bye' are the action at street corners. Just as Lefebvre suggests, 'through the mediation of [multiple] rhythms, an animated space comes into being' (1991: 207).

Beyond the pulse of movement, rhythmanalysis 'does not neglect smell, scents, the impressions […] which society atrophies' (Lefebvre, 2004: 21), thus, attending to sights sounds and odors reveals their measures too. Walking past a deli, I hear a splash of Polish, or maybe it's Latvian or Lithuanian. I catch a waft of salted fish and an eyeful of yams from a shop that sells African provisions. Inside, a group of women chitchat in Yoruba, while a television, off in the corner, beams images and sounds from West Africa to this here and now. An employee, 'planted' outside a travel agency, hands out flyers advertising deals to destinations for folk who might want a break and a different pace. There's a hair-braiding outfit, an international call centre and a language school and the plethora of posters announcing various 'inter-cultural' extravaganzas suggest a cacophony of diversity in the social spaces around Ireland today.

Each spatial and social act – including those associated with consumption, exchange, proximity, distance, communication and desire – has a rhythm that brings to this scene 'a multiplicity of sensorial and significant meanings' (Lefebvre, 2004: 32). At the same time, the prevailing beat, the downbeat, is one that bespeaks novelty, diversity and change; to use Lefebvre's words, 'it screams, "Down with the past! Long live the modern! Down with history"' (32). These ethnographic observations also resonate with social scientific accounts describing such newly heterogeneous 'ethno-cities' as manifestations of the rapid redefinition of Irish society (see Corcoran, 2006; Sparks, 2006).

As I became ensnared in the dominant beats of Ireland's hasty 'progress' and gained a sense of the tempo of this more cosmopolitan capitalism, I found my thoughts arrested by what has been described as a 'perennial Yeatsian soundbite' (Harrington, 2005: 428) from the poem 'Easter 1916' which W.B. Yeats wrote at another moment of tumultuous change in Ireland, and where a line repeats 'all is changed, changed utterly, a terrible beauty is born' (Yeats, 1970 [1921]). In reciting Yeats here, I am not – as some interpretations of Yeats' works suggest – calling for a return to some mythically imagined past; instead, and more in keeping with my interest in the rhythmic interplay between change and repetition, I rephrase Yeats' lyrics as a question and ask, is it utterly changed after all?

To the offbeat: rhythms of waiting

Analyzing the profusion of rhythms on a coach tour of the Ring of Kerry, Edensor and Holloway (2008) observe the selective representation and retelling, and associated fluid temporalities, of history and the present. They note that the growth and success of Ireland's tourist industry turns on the marketing of Celtic history and tradition, rustic landscapes and folksy people, where a 'highly selective, orthodox tourist production […suggests] nought of the dynamic economic success of the emergent "Celtic Tiger"' (2008: 489) and little of other post-nationalist developments in politicsm and culture. Beyond calling attention to the diffuse rhythms that hum alongside dominant beats, they also highlight how old and new recur as requirements for success in the contemporary heritage industry.

My excursion through small-town Ireland invites further insight into the dynamic interplay between past and present that inheres in the rhythms of everyday life. In place of the endeavors of the tourist industry to recapture Ireland's past, my ethnographic encounter picked up the 'Celtic Tiger' rhythms where innovation, connectivity, diversity, and newness have shaped the dominant beats. But, coinciding with these scenes of change that herald Ireland's rapid transformation there are also offbeats where 'the older, the archaeo, resurfaces' (Lefebvre, 2004: 52). By calling attention to these offbeats, or by taking up a syncopated beat, we can highlight how everyday rhythms complicate chronological orderings of past and present and how what appears 'utterly changed' repeats in fascinating ways.

Shifting the emphasis then, from the dominant beat to the offbeat, I want to turn attention to the end point of my journey and to the narratives of day-to-day life recounted by a group of women, most of whom had arrived in Ireland from Nigeria and were waiting for their applications for asylum to be reviewed. As previously mentioned, the Irish government operates a policy of housing dispersal and 'direct provision' for asylum seekers. As soon as requests for asylum are filed, applicants are bussed to one of 64 accommodation centres dotted throughout the country, where they are provided with room and board facilities while they wait for a decision on their application for refugee status.

One of the centres I visited was a former convent, which, with declining numbers of women entering religious vocations, was being used as a temporary home for asylum seekers. This converted convent was perched on the outskirts of another small town; not unlike my midland stop, this town had a main street where shifting demographics, tastes and lifestyles were evidenced in the multi-lingual signs of supermarket awnings, unusual fare in shop windows and ads for English classes in the library. But as I conversed with several of the women living in the town's newest enterprise – the accommodation centre – their accounts of everyday life suggest a syncopated rhythm to coincide with these polyrhythmic beats. Within the hollowed hallows of the former convent and amidst what Lefebvre describes as the 'ocular and verbal chatter' (2004: 47) of the everyday – of babies crying, children laughing, the meter and laughs of a Gaelic football game and dashing inside with the threat of rain – and in between the structured schedule of the accommodation centre, these women revealed a day-to-day pulse of a different order.

In this place, residents' everyday lives are draped in the schedules and rhythms of linear time as a brief recounting of their daily routine illustrates. First thing in the morning they must sign-in with hostel authorities. The system of 'direct provision' means that meals are provided cafeteria-style and residents are not permitted to prepare their own food; consequently, breakfast is served in a common dining room at 8:00 a.m. After breakfast, women walk their children to school, which starts at 9:00 a.m. Then, some attend a morning church service in the town and return to the accommodation centre by 10:00 a.m. After that, and in the absence of a productive place or role, these women spend much of the rest of the day caught up in a humdrum rhythm of sit and wait. While their applications are being reviewed, asylum seekers are not permitted to work, they can partake of educational programs, though in very limited ways, and a reliance on public transportation means that opportunities to travel beyond the town are costly and infrequent. Although the speed with which applications are processed has improved in recent years when this study was completed, in 2004, processing took anywhere from nine months to two years.

Instead of being part of the multiplicity of fast-paced rhythms of Ireland's recent economic successes, these women are relegated to the sidelines of the 'Celtic Tiger' upswing. For them, the everyday is embodied as a seemingly interminable encounter with immobility, boredom and suspended time. In conversations I invited these women to convey a sense of their day-to-day experiences before and upon becoming an asylum seeker in Ireland. One woman, who had been a medical doctor in West Africa before arriving in Ireland some 13 months prior, described her day-to-day life as follows:

> [My life before…] that is a faraway dream. I was working as a registrar, I was working in one of the biggest hospitals in West Africa […] it means a very, very tight schedule […] always busy, always busy.
>
> [DC: And now?]
>
> I sit here doing nothing, [….]
>
> I can't prepare a meal, to get milk, to get bread, I have to ask [….]
>
> So, when the children come home from school, we go to the kitchen, we eat dinner, come back here, I take their clothes to the laundry, [and] basically that's my life. (Interview with Sandra, 30 October, 2004) (Interview participants' names were changed at their request in order to protect their privacy)

Several women conveyed a sense of frustration related to their perceived marginal status, the protracted state of limbo they encountered, and the way these experiences impacted the rhythms of the day-to-day. As one participant explained:

> I don't have the right to study, I don't have the right to work; basically I can't do anything other than sit around. (Interview with Irene, 30 October, 2004)

While another woman expressed this more poetically, noting:

> Living here, I feel like a caged bird that just wants to be free. (Interview with Lisa, 30 October, 2004)

Yet another woman explained that while she was happy that she now had some peace of mind, the banality of day-to-day life made her weary. In an effort to alleviate the boredom she recounted that she had tried to take up knitting, an old craft tradition that she associated with Irish women, but such sedentary tasks and the passive routines with which she filled her time did not relieve the sluggish pace of daily life. With this, she conveyed dismay at how these rhythms were quite literally embodied:

> The only thing I don't really like about myself here, about my life now is not working, and being idle all the time. I just want to do something, even voluntary,

just keeping busy, look at me, I'm getting fatter; I wasn't like this when I came.
(Interview with Beji, 30 October, 2004)

This sense of inactivity, ennui and unease in the face of these slow rhythms was
repeatedly echoed in these stories of everyday life. What is striking about these
accounts is their irregularity when juxtaposed with the pace and tempo that are
etched out in the dominant rhythms and accounts of Ireland's recent dash to
global post-modernity. For these women, and for others seeking asylum during
the 'Celtic Tiger' years, the everyday was certainly fixed and ordered within linear
rhythms of production and consumption, but coinciding with this, their day-to-day
lives were embodied as the slow-paced measure of idle time. Taken together, the
regular and syncopated rhythms detailed here highlight the concurrent assemblage
of measures that unfold within the ordinary social spaces of the everyday.

A polyrhythmic chorus that's changed but nor *utterly* changed after all

The syncopated measures marking asylum seekers' everyday lives also echo older
and quite familiar beats within Irish society. The rhythms of waiting and idleness
that the women describe were also the rhythms of large-scale unemployment
and 'staying put' in rural places when emigration was the order of the day (see
Lee, 1989; O'Connell, 2001; Delaney, 2004). Drawing on the oral histories of
individuals who, for one reason or another, did not emigrate from Ireland in earlier
eras, Gray (2003; in prep) focuses attention on experiences of remaining at 'home'
during periods of massive out-migration such as in the 1950s; in doing so, she
highlights efforts to 'harmonize progressive mechanical time and the lived sense
of enduring time' (in prep: 4). Gray details stories of waiting and anticipation;
waiting for word of work, for letters and remittances and for the occasional holiday
visit. These beats occurred in conjunction with anticipating chances to leave and
the possibilities of a sibling, friend, or lover's return. Gray describes the non-
migrant's everyday life as one immersed within a 'culture of waiting' (11). In the
routines and syncopated rhythms of asylum in contemporary Ireland, this culture
remains alive and well.

Moran observes that 'the everyday complicate[s] any notion of the past as dead
and buried, disconnected from the present' (2004: 66). In my view, the slow-paced
rhythms of day-to-day life described here work to further ground this claim. As the
practices of the past coincide with the rhythms of the present, these reverberating
beats interrupt the linear and progressive ordering of history to which we are
accustomed. In addition, another element of Lefebvre's sophisticated analysis of
everyday rhythms is highlighted here. While the day-to-day accounts I've detailed
demonstrate that the past recurs amidst routine rhythms of waiting in the present,
this apparently bygone but now proximal social and spatial practice today affects
a distinct social body, that of the asylum seeker. In this sense, these repetitions
cannot be apprehended as exact reproductions of historical time and space but

instead, as Elden points out, there is the paradox of 'the generation of difference [in and] through repetition' (2004b: 179). As the bodies of asylum seekers replace the unemployed and other 'remainders' from earlier eras and as changed social spaces crisscross these continuities and discontinuities the 'differences [that are] induced or produced by repetitions constitute the thread of time' (Lefebvre, 2004: 8) as it unfolds in social space.

To review, the downbeat or dominant rhythms of everyday life observed in these scenes from contemporary Irish society might suggest that all is changed. But the syncopated rhythms of waiting and ennui in asylum seekers' accounts serve as a corrective to 'narratives of abundance and innovation' (Moran, 2004: 55) by revealing the traces of the past in the present. In tandem with this, the beats that keep time with the past also 'renew themselves' (Lefebvre, 2004: 33) producing a dynamic polyrhythmic chorus. Overall then, we might think of Lefebvre's project on rhythmanalysis as a symphony of several movements: a prelude comprised by a crescendo in the critique of modern life, an overture of polyphonic and syncopated tempos of social space, and a recapitulation where continuity, discontinuity and differences inhere, and all of which unfolds amidst the fascinating rhythms and echoes of the everyday.

Chapter 7

'I'm in a Park and I'm Practically Dead': Insomnia, Arrhythmia and Withnail and I

Craig Meadows

Introduction: *Withnail and I* and insomnia

Locked in a sixty-hour spell of unbroken wakefulness and drug-induced delirium, and horrified by the everyday regime of urban life in London, Marwood, in the 1987 film *Withnail and I*, opines that his 'heart is beating like a fucked clock.' For Marwood, even a broken clock tells the right time twice per day, but his arrhythmic heart is incapable of achieving even the kind of minimalistic exactitude found in paralysis. As such, Marwood's insomniac existence is defined by a general drift towards what he calls the 'realm of the unwell.' This suffering is marked by a two-fold exclusion. The first is from the presumably 'natural' circadian rhythms of sleep and wakefulness. The second exclusion is from the repetitive structure and institutions of everyday life. Marwood and Withnail thus represent figurative starting points for a rhythmanalysis of insomnia as well as a critical reading of Henri Lefebvre's conception of arrhythmia (2004).

Chronic insomnia entails a protracted inability to fall asleep or sustain restful sleep. The insomniac body undergoes a variety of physiological stresses stemming, in part, from increased levels of hormones that further erode the possibility of drifting off into sleep. These physiological stresses conjoin with an accumulation of social, economic and psychological stresses and anxieties, and this totality is directly related to mutually reinforcing conditions such as substance abuse. The result is often a spiralling inability to sleep, such that the hours of wakefulness extend beyond the confines of the twenty-four hour day and sleep, however fitful it may be, lasts for no more than an hour or two before another prolonged period of wakefulness begins. Even when sleep is attained, the insomniac does not descend into deeper stages of sleep like others. Prolonged wakefulness of thirty hours or more creates sensations that include burning limbs, blurred vision, intense cranial pressure as well as a fugue-like state, and eventually states of mania. This accumulation of wakefulness produces a discordant body at odds with circadian rhythms and the repetitive patterns of everyday life. Extremely long periods without rest or years of suffering culminate in dementia or despair, thereby associating insomnia, somewhat problematically, with depression. The production of despair in insomnia leads to the sense that it is a condition unto death, yet, with the exception of the extraordinarily rare case of Fatal Familial Insomnia, death is the one thing that insomnia refuses to deliver. Under these conditions, it is no surprise that extant narrations of insomnia focus upon dementia and social pathology. What rhythmanalysis, in conjunction with *Withnail and I*, opens up, however, is a possibility of reading insomnia as a tension between the critical capacities contained within the twenty-four hour watchfulness of Lefebvre's 'philosopher' and the definition of arrhythmia as a pathological state in which the accumulation of this watchfulness in the subject collapses into nothingness.

I begin with a discussion of Lefebvre's (2004) notions of cyclical and linear rhythms and then moves into an exposition of the film *Withnail and I* in order to situate arrhythmia in relation to established understandings of rhythm. I then shift to a discussion of the relation between cosmic/circadian rhythms and the social construction of normed sleep in both Lefebvre and the sleep sciences, with particular reference to urban space and the institutions of everyday life. Thirdly, I look at Lefebvre's notion of dressage and modify it with reference to Adorno's conception of the 'organic composition of capital' and the social patheogenesis of schizophrenia. Finally, I look at the role of parks and their purported value as a restorative for the urban dweller and the critique of this concept offered in the film. I conclude with reference to Lefebvre's notion of the twenty-four hour philosopher, the critical capacities of insomnia and the accumulation of wakefulness.

Arrhythmia, insomnia and Lefebvre's everyday life

The insomniac emerges in two moments in *Rhythmanalysis*. The first, in 'Seen from a Window,' depicts the insomniac as an apparition that turns on a light in a window in the desertified realm of nocturnal urban existence. This moment is otherwise marked only by the ongoing march of traffic signals that impose signs of social regulation on the void that surfaces from the dissolution of daytime rhythms (Lefebvre, 2004). Lefebvre (2004) also posits insomnia as a methodological tool that, once introduced, threatens the regime of everyday life with collapse. Both of these instances establish an equation of insomnia with arrhythmia. Firstly, insomnia is the upsurge of an unexplainable, dream-like, solitary and unregulated spirit from within the void on which social regulation, in the form of the traffic signals, attempts to leave its mark. In the second, and corollary instance, insomnia is an outpouring from this void that escapes social regulation and creates what Lefebvre calls 'another everydayness,' (2004: 75) which he then dissolves in ellipses. This other everydayness is rendered invisible and therefore incomprehensible. This section thus explores this subsumption of insomnia within the conceptual analysis of arrhythmia and analyses Lefebvre's definition of arrhythmia as a pathological state.

Rhythm, as the basis of everyday life, is the combination of linear movements and cyclical repetitions, which bring together the mechanistic, rational and structural elements of social life, leisure and work with cosmic or organic cycles. The cosmic is manifested most clearly in circadian rhythms of night and day, in the production of what Lefebvre calls 'differentiated time' (2004: 78). This differentiated time is marked by repetitions that produce variations and effectively mark off measures of time. However, in the process of bringing these terms together in a dialectic unity, Lefebvre seemingly conflates his definitions. In 'The Critique of the Thing,' Lefebvre begins by arguing that the relationship between the cyclical and the linear is one of an antagonistic unity:

> Cyclical repetition and the linear repetitive separate out under analysis, but in *reality* interfere with one another constantly. The cyclical originates in the cosmic, in nature: days, nights ... monthly cycles, etc. The linear would come rather from social practice, therefore from human activity: the monotony of actions and of movements, imposed structures. (2004: 76)

However, in a later analysis he adds another understanding of the cyclical by equating it with the social and dissipating its connection to nature. In 'Seen from the Window,' Lefebvre argues that recurring rhythms, such as the daily appearance of shoppers or schoolchildren at certain times of the day, compose cycles:

> [O]f large and simple intervals, at the heart of livelier, alternating rhythms ... The interaction of diverse, repetitive and different rhythms animates ... the street and the neighbourhood. The linear ... consists of journeys to and fro: it combines with the cyclical, the movements of long intervals. The cyclical is social organisation manifesting itself. The linear is the daily grind, the routine, therefore the perpetual, made up of chance and encounters. (2004: 30)

In this latter exposition, social organisation shifts from the linear to the cyclical. His notion of *linear repetition* is at the heart of this shift, in which the cyclical takes on properties of social organisation. The linear forms repetitions through its infusion with the cyclical, such that hourly demands and social practices compose rhythmic regimes in their repetition across days, weeks or seasons. To reiterate, there are two distinct uses of cyclical at work in Lefebvre's text; firstly, that which comprises cosmic repetitions such as circadian rhythms, and secondly, that which captures the repetitive structure of linear tasks of everyday life. This connection of the cyclical and social organisation occurs on the plane of linear repetition, where he fuses the cyclical and linear together. In Lefebvre's second volume of the *Critique of Everyday Life,* he adds another layer to this distinction. He connects the cyclical to the cosmic and then argues that the cyclical has been subordinated to the linear time scales of social life (2002: 48). He further contends that customs that have been historically linked to cyclical time are 'deeply rooted' and persist in their cosmic form despite 'anti-nature' attempts toward emancipation from these rhythms (2002: 48-9). Thus, the cyclical has the capacity to organise daily life through its reference to the cosmic, while it is also increasingly determined by linear, rational methods of social organisation.

Ultimately, Lefebvre's definition of rhythm separates out a dialectical unity of cyclical and linear movements, as well as repetitions and ruptures (what he refers to as 'moments') that are discernible in the structures and patterns of everyday life. Alternations between diurnal and nocturnal rhythms conjoin with the simultaneously random and organised tasks of work, leisure, motion and rest. More specifically, he argues that the body is the organic focal point of these regimes, of a 'regulated time, governed by rational laws' that 'impose themselves on the multiple *natural* rhythms of the body ... though not without changing them' (2004: 9). As such, Marwood's arrhythmic heart, and thereby the experience of insomnia, becomes the focal point of understanding how regulated, rational patterns, 'natural' circadian rhythms and movement through urban space interact to produce insomnia as an arrhythmic accumulation of wakefulness. This also raises the question of how 'natural' circadian rhythms can be linked to experiences such as sleep. The method of this analysis follows Lefebvre's call to 'unwrap the bundle' of rhythms that compose this body; the twist, however, is that Marwood and Withnail's bodies are, at their cores, the negation of rhythm.

Arrhythmia as supplement to everyday life: Marwood's flight to the diner

Withnail and I begins with Marwood leaving his flat at dawn for breakfast at a local diner. As Marwood broods over his meal, he watches as greasy eggs are slapped between slices of white bread and then consumed by a woman across from him. As the egg yolk oozes from between the slices of spongy-looking bread, Marwood retreats in horror behind the daily paper. However, Marwood's gaze turns to a newspaper headline, 'Love made up my mind: I had to become a woman,' and a look of panic enters his eyes. Marwood thinks to himself, 'Thirteen million Londoners have to cope with this, and baked beans and All-bran and rape? And I'm sitting in this bloody shack and I can't cope with Withnail' (Robinson, 1987). There is no reprieve for Marwood: the external world offers horrors and his dwelling place, shared with Withnail, is oppressive. This oppressiveness is most clearly articulated in a subsequent scene when the flatmates attempt a foray into the kitchen in order to tackle the accumulated detritus and rodents that emanate from the sink. What had begun as a near flight from his flat thereby culminates in his failure to integrate into the institutions of everyday life.

Despite its position as part of a discernible rhythm of daily life, the diner does not represent a eurhythmic space for Marwood. Eurhythmia occurs when rhythms 'unite with one another in the state of health, in normal (which is to say normed!) everydayness; when they are discordant, there is suffering, a pathological state' (Lefebvre, 2004: 16). In a eurhythmic state, the movement across differentially segmented portions of the day (the movement from the time of sleep in the dwelling to the entry into the world and the institution of the diner/the newspapers) would be experienced as a smooth transition across functionally organised spaces and times. Entry into the diner and the mediatised day of the newspaper would constitute part of a putatively ordered and productive existence. The diner represents an urban institution that produces everyday life, yet, despite its role in the diurnal rhythms of city life; the diner simultaneously entrenches Marwood in his arrhythmic state and leaves him pondering how anyone can function according to these socially organised, yet antagonistic rhythms. Marwood's perceptual experience of the diner cannot synthesise these rhythms.

His perception points to a negativity at the heart of the production of everyday life, in that the diner renders eurhythmia impossible even as it produces the rhythms of everyday life. Marwood's perception is of a discordant world and his experience is apparently pathological, as he is incapable of functioning in this world, and is in concord with Lefebvre's definition of arrhythmia, in which discord produces suffering and a pathological state. However, there is a significant shift in an understanding of pathology here, for Marwood's experience shifts the discourse of insomnia away from diagnostic pathologies of cause and their relation to the 'symptoms' of insomnia. His reactions constitute an alternative assessment of the regulated rhythms of daily life: these rhythms proffer a productive, urban existence, but are simultaneously incapable of assimilating the arrhythmic subject (Marwood). This subject is itself a symptomatic presence that undermines the ostensible naturalness of these regulated rhythms.

Lefebvre also argues that the media, manifest in the paper that Marwood ducks behind, produces a deceptive fullness in this regime. The media represents the possibility of understanding the everyday, but instead offers a simulacrum of 'empty words' and 'mute images ... a present without presence' (2004: 47), which is the horror from which Marwood retreats. While a full critique of the fetish of presence and the poetic in strands of modernist thought is beyond the scope of this essay, Lefebvre's critique of the mediatised day dovetails with Marwood's musings on the way in which rape and All Bran enter the flow of everyday life as information, thus 'fabricating, introducing and making accepted the everyday' (Lefebvre, 2004: 47). The dominant position of the media stalks the intrepid explorer, Marwood, but arrhythmia, as a method, denies the legitimating role that the media plays in the repetitions of everyday life. Instead of banal banter on the media-produced present, this manufactured present torments him to the point of hysteria.

It is a matter of pointless speculation to ask whether the other figures in this scene live this experience as a eurhythmic state. Nor is this a matter of a universe of polyrhythms, wherein multiple, but not necessarily harmonised, rhythms co-exist and the subject is a self-reflexive constituent of modernity. Arrhythmia, like the Derridean supplement, cannot be incorporated into the 'whole' as a complementary existence. For Derrida, the supplement 'intervenes or insinuates itself *in-the-place*-of ... If it represents and makes an image it is by the *anterior* default of a presence ... As substitute, it is not simply added to the positivity of a presence, it produces no relief' (1976: 145). Marwood skims across the surface of the everyday, experiencing it in its unbearably heavy emptiness and unable to immerse himself in its rhythms.

Circadian rhythms, normed sleep and urban space

Of particular significance to the analysis of arrhythmia is Lefebvre's positioning
of this 'fatal disorder' as a void that reigns over urban space between the hours of
3 a.m. and 4 a.m. For Lefebvre, the penetration of diurnal rhythms into the night
represents an accumulation of the activities of wakeful life (the ordered regime of
capital). However, these displaced diurnal rhythms are simultaneously transformed
and their pace is slowed to the point that they often dissolve. The insomniac, in
turn, emerges as an unregulated figure in this transformed space. In 'Seen from the
Window,' Lefebvre observes, '(T)he night does not interrupt the diurnal rhythms
but modifies them, and above all slows them down … [Yet at o]ther times, there
is no-one at the lights, with their alternating flashes … and the signal continues to
function in the void … before the façades that dramatically proclaim their vocation
as ruins' (2004: 30). There are two important aspects to this passage. The first is the
way in which daytime rhythms persist into the night, only to eventually dissolve,
which Lefebvre evokes in a manner similar to Benjamin's project of uncovering
modernity as ruins. Moreover, in summing up this 'scene', Lefebvre asserts that,
'arrhythmia reigns, except for rare moments and circumstances' (2004: 31). Thus,
while Lefebvre does not assign a cause, symptom or effect to arrhythmia, he does
assign it its own place and time: at four o'clock in the morning in the inactivity of
the streets and the light in the window. This is the darkness from whence Marwood
emerges at the beginning of the film. Marwood's arrhythmic appearance on the
streets is thus as a hostile entity that does not share in the 'miraculous charm'
(Lefebvre, 2004: 73) of the advent of dawn and the restoration of diurnal rhythms.
This situation of arrhythmia places it as the negation of the circadian rhythm,
understood as a cycle of sleep and wakefulness 'naturally' linked to the cycles of
light and dark, activity and inactivity.

The fundamental cosmic rhythm that modern, capitalist society associates with
sleep is that of the circadian alternations between day and night, which pivots
on the notions of falling asleep (in the evening) and awakening (at dawn). Other
attributes have been interpellated into this rhythm, such as the Enlightenment
project's situation of knowledge and consciousness within the day, light and
awakening. The premise of the circadian rhythm, whether it is associated with a
natural cosmic or socially determined order, is that the body accumulates a 'sleep
need' through the progression of the day. While this need accumulates, the presence
of the light of day suppresses the urge to sleep. With the onset of night, the body
succumbs to sleep and dispels its accumulated need, while darkness holds the
sleeper in this state, despite a steadily decreasing sleep need. This understanding
of the circadian rhythm posits a an image of a homeostatic being capable of an
optimal functional existence. Kenton Kroker (2007) notes that this functionally
positive and rhythmic understanding of sleep emerged around 1900. In the mid-
twentieth century, with the emergence of EEG analysis, a method of recording
electrical activity in the scalp to ascertain neuron activity in the brain, the block of
sleep was assigned its own temporal architecture through the separation of periodic

alternations between different wave forms (Kroker, 2007). These cycles within the cycle of sleep/wake alternations serve to deepen the notion of not only sleep/wake cycles but also advances notions of a linear repetition of approximately two-hour blocks of sleep.

The first question concerning the association of circadian rhythms with the sleep/wake cycle is how this is produced as a putatively natural rhythm. While sleep science has developed a shifting topography of explanations for a biologically based, functional explanation of sleep, the conception of the uninterrupted block of nocturnal sleep remains at the core of approaches to insomnia. This is the basis of the psychopharmacological approach, which seeks to medically alter the levels of serotonin in the brain through the use of anti-depressant, SSRI (Selective Serotonin Reuptake Inhibitor) drug regimes. Serotonin, according to the theories of the sleep sciences and pharmaceutical corporations, is a brain chemical that purportedly regulates moods and the ability to sleep. When prescribed to insomniacs, SSRI drug regimes are supposed to help increase what is surmised to be a damaged or permanently altered level of serotonin in the brain. These drugs purportedly harmonise the sleep patterns of the subject with this naturalised notion of sleep cycles.

Non-pharmacological approaches to insomnia tend to focus on 'sleep hygiene', where insomniacs are blamed for their condition on the premise that insomnia is a product of accumulated bad sleep habits. Sleep hygiene ranges from telling the sufferer to avoid substances such as caffeine, chocolate and strenuous exercise in the evening and establishing set wake-up times in the morning. This will supposedly establish normed sleep patterns. Yet sleep hygiene is wholly inadequate for figures such as Marwood and Withnail, who are completely dislodged in time. To prescribe a set wake-up time to someone living beyond the boundaries of the 24-hour day is patently absurd. Ultimately, sleep hygiene is a cynical ornamentation applied to the regime of sleep in modernity, a type of ornamentation that Lefebvre scathingly critiques. For Lefebvre, capital 'erects itself on a contempt for life and from this foundation: the body, the time of living' (2004: 51). Lefebvre argues that this contempt manifests itself in an ethic and mobilises ornamentation to provide cover for its contempt. The practice of blaming the insomniac is a clear example of this sleep related 'ethic'. The definition of chronic insomnia places fault for the condition squarely upon the sufferer, as do various cultural conceptions of the insomniac that are always reducible to the notion of an immoral person unable to moderate his or her thoughts and practices.

One of the key conclusions concerning circadian rhythms comes from the sleep researcher Nathaniel Kleitman in the 1920s. He argues that rhythm is a conditioned response that is individually acquired and depends on an extrinsic reinforcement for its establishment, yet will persist for a period of time in the absence of this reinforcement (Kroker, 2007: 307). While Kleitman argues that this consolidation is part of an evolutionary process that produces benefits for the species, his assertion does de-articulate the naturalised overlaying of sleep/wake cycles onto circadian rhythms. Prior to the emergence of insomnia as a pathological condition in the 1880s, sleeplessness was a pre-modern term used to describe periods of wakefulness between what were known as first and second sleeps. This highlights how the consolidation of the eight hour, nocturnal sleep block in modernity, like the sleep blocks of pre-modern life, is part of the social organisation of life and work. Kleitman's assertion that sleep is a product of enculturation is therefore another way of saying that sleep is a performed social act. This argument works as a parallel to Lefebvre's assertions concerning the determining role of social organisation and the need to immerse oneself in rhythms to restore states of good health. Understanding sleep as social performance on these terms necessitates a methodology of restoration of the conditions of performance that produced the eight hour block in the first place (Kroker, 2007), while the purpose of my work is to analyse the tensions at work between this regime of sleep and wakefulness and insomniac existence.

Dressage: capital, nature and sleep

In the *Grundrisse,* Karl Marx argues that the worker requires ten to twelve hours of rest in order to sufficiently restore the capacity to repeat labours and exchanges with capital (Marx, 1973). While I can only speculate as to the veracity of such sleep quantities as an actual experience of industrial labour (did the labourers depicted in Engel's text *The Condition of the Working Class in England* get this much sleep?), Marx points to the way in which sleep was subordinated to the emergence of industrial labour and Marx's notion of ten to twelve hours of rest puts into relief the dwindling expectations concerning sleep quantity over the past 150 years. While sleep itself may be 'natural' or necessary, it is an open question as to the relationship between 'nature' and how individuals sleep within advanced capitalist societies. As I have already noted, Lefebvre understands the relationship between cosmic and linear rhythms as one of interpenetration, with socially determined linear repetitions increasingly defining cosmic rhythms. Yet Lefebvre retains the proviso that the regime and institutions of everyday life remain 'shot through and traversed by great cosmic and vital rhythms: day and night ... and still more precisely biological rhythms' (2004: 73). While Lefebvre never explicitly connects sleep/wake cycles to these cosmic rhythms, the implication of his narrations and conceptualisations of cosmic rhythms is such that sleep cycles are 'shot through' with cosmic rhythms and thereby found a biologically determined circadian sleep rhythm. The sleep regime in capital is thus recuperated as being at least 'of nature'. What this thinking achieves is a restoration of biological thinking, and with it a thinking of a nature that founds social life and is yet deformed by it.

Yet, Lefebvre also argues that these cosmic cycles are subordinated to circadian rhythms of urban existence. This is done insofar as the sleep cycle is itself a deformation of some more cosmic mode of being. Lefebvre suggests that the everyday is shaped by the homogenous times and measures of work and the subsequent transformation of cosmic rhythms by the rational laws that imprint themselves upon cyclical patterns, to which the various activities of everyday life are subordinated. This notion of deformation is a tricky affair. Lefebvre (2004) argues that capital is anti-nature (a logic as questionable as that which posits capital as nature) and he also argues that cosmic rhythms are deformed by rational laws. Thus, Lefebvre presents us with an image of circadian rhythms that are socially constructed patterns opposed to nature, yet shot through with and founded upon vestiges of nature. Yet both elements of this image are problematic. The notion of opposition to nature undermines the organic aspects of social life, and moreover, it is impossible to validate the speculative claim that a retained cosmic nature may inform sleep patterns.

This necessitates a separation of this strand of thought from the otherwise productive notion of dressage. He defines dressage as a 'bending' by which the subject is broken in, through a series of repetitions of linearly organised 'imperatives and gestures' (2004: 39). Lefebvre posits the basic structure of this regime of sleep in the tripartite division of dressage into times of 'internal activity of control... integral repose (sleep, siesta, dead time) ... [and] diversions and distractions' (2004: 41). This division of the time of dressage carries with it a curious, but significant parallel to the modernist, urban functionalism of the CIAM (Congrés International d'Architecture Moderne, founded 1928) plan for the functional city, which divided the city into spaces of work, residence and leisure. Like CIAM which assigns a functional space for sleep, modern sleep science conceives of sleep in terms of functionality and time. Lefebvre draws on these divisions in his functionalised typologies of dressage. All three render sleep as a uniquely defined activity, which occurs in a specific place (within the functional dwelling in the topography of the modernist city) and in a specifically assigned block of time (the time of repose). By implication, the disciplinary regime of modernity takes shape in the division of time within the body and the division of space across the city, and that these divisions attempt to fuse together the time and space of sleep. In the most basic sense, *Withnail and I* shows the replication of the logic of spatial divisions between dwelling, diner and park and the concomitant assumptions of the functional roles these places have, while the arrhythmic existence of its characters undoes these assumptions.

The question, however, is how to understand the relationship between regimes of sleep and wakefulness and circadian rhythms, which Lefebvre explicitly critiques and yet seemingly integrates into his narration of arrhythmia and in his retention of a nature/anti-nature divide. Theodor Adorno offers a slightly different terminology and approach to this issue. He uses the term 'organic composition of capital' to point out the 'growth in the mass of the means of production,' which 'designates subjects more and more exclusively as partial moments in the network of material production' (2005: 229). This basic claim of the functional objectification of subjects in capital is expanded upon when he argues that this is no mere mechanisation of man. This is not just 'specialized technical faculties, but ... equally their opposite, the moments of naturalness which once themselves sprung from the social dialectic and are now succumbing to it ... causes man to pay for his increasing inner organization with increasing disintegration' (Adorno, 2005: 229-31). Adorno challenges notions of a deformative, external influence with the notion that modern society produces an organicised 'hereditary taint' that biologism in turn projects as a component part of nature. The organic composition of capital offers no recourse to a natural cosmic rhythm within the order of capital. Seemingly natural patterns, such as circadian rhythms, are, in their totality, a product of socially organised rhythms, but the unhealthy and overlapping nature of these multiple, constitutive rhythms serves to undo their organic spontaneity. In other words, the structural organisation of everyday life in modernity collapses into arrhythmic states not strictly because of some conceived deformation of an underlying nature, a 'bending' that goes too far, but because the tensions that structure our movements through everyday life produce social and individual disintegration alongside the naturalness that they initially produce. Adorno calls this process the social pathogenesis of schizophrenia. As with Lefebvre, there is no linear cause/symptom progression in Adorno's schizophrenic conception, rather arrhythmia emerges out of the seemingly organic, organised rhythms of everyday life.

Adorno's conception of the organic composition of capital and the social pathogenesis of schizophrenia clarifies the relationships between the subject, social organisation and cyclical patterns. Instead of an imprint on a vestigial remnant of a different rhythmic order, Adorno points to how adaptation to the rhythms of capital conjoins with an internalised, intuitive naturalness. Linear repetition is thus not 'shot through' with cosmic cycles; rather it is validated by reference to these cycles. The point here is not to say that there is no such thing as a cosmic rhythm, but that there is no possibility of discerning some supposedly 'natural' wakeful/ sleeping rhythm that could be organically mapped onto these cosmic rhythms. Lefebvre is similarly critical of the social production of everyday life, but the notion of a cosmic ground of bodily experience is problematic.

Nature and Regent's Park: Marwood's search for harmonic restoration

After the aborted attempt to clean the kitchen in their highly toxic flat, Withnail and Marwood leave the house for Regent's Park, where Marwood suggests that what they need is to escape to the country in order to rejuvenate. Withnail sardonically replies, 'Rejuvenate? I'm in a park and I'm practically dead. What good's the countryside?' (1987). Withnail's comment undercuts the traditional Anglo conception of nature as a cure-all for purportedly urban ailments, which is underscored in the park and by the mayhem that ensues upon their arrival in the countryside. While the countryside can offer an escape from the intrusive presence of artificial lighting, which is a precondition for the extension of diurnal rhythms into the night, even grand scale parks and the countryside cannot deliver the promise assigned to them. Frederick Law Olmstead aptly summarises the dominant cultural understanding of the park when he lovingly surmises, '(T)he enjoyment of scenery employs the mind without fatigue and yet exercises it, tranquilises it and yet enlivens it; and thus, through the influence of the mind over the body, gives the effect of refreshing rest and reinvigoration of the whole system' (Rybczynski, 1999: 258). This supposedly functionalised spatial harmony of the park is undone by Withnail's declaration, since his accumulation of sleep loss manifests itself as a fatigue that cannot be restored by the paltry offerings of this park-based hygiene. While Lefebvre (2004) might critique this method of importing the cosmic into the linear as yet another example of dressage and of the contempt that capital has for the individual in its mobilisation of hygiene, the narrative of *Withnail and I* explicitly critiques his notion, as well as that of the sleep sciences and psychopharmacology, of restoration through re-immersion in rhythms. As most any insomniac could confirm, the quest for a restorative rhythm is the illusory image of eurhythmia that always eludes one's grasp.

Lefebvre offers a more thorough critique of the park, and ultimately the countryside as well. The park is offered as a sort of still life or homeostatic image that purportedly harmonises other modes of urban existence through reference to nature. Marwood outlines his quest on these terms. He desires to restore a rhythmic existence through inhabiting the landscape, which he conceives in terms of nature and further escape from the horrors of everyday life in London. However, Lefebvre (2004) critiques this conception of the space of the park as a sort of vestige or parody of nature for it fails to take into consideration the fact that each image or succession of images of the park or country hides a polyrhythmic world: still life oscillates. Lefebvre's critique has two levels of meaning. Firstly, the park is not nature and cannot restore a cosmic rhythm. The park is a modern social institution that is proffered as a harmonising and naturalising influence, but is actually a parody of nature and a deepening of the functionalist logic of urban space. Secondly, our image of the park is caught up in notions of landscape and still life imagery, which elide the oscillations contained therein. Ultimately, the park cannot deliver 'nature' as conceived by Olmsted and it is therefore unable to restore a rhythmic existence to the suffering insomniac, including Olmsted's own lifelong chronic insomnia.

Yet, there is a fundamental difference between Lefebvre's articulation of the polyrhythmia inherent to the still life and the experience of Marwood and Withnail. Lefebvre's excited narration of the rhythmic existence beneath the still life echoes his articulation of the wondrous charm of the dawn. What is lacking in this celebration of rhythms is a full exposition of the manner in which these rhythms are constituent elements of modernity, as is the case with Regent's Park. Marwood and Withnail are unable to articulate a harmonic relationship between themselves and the park. But their arrhythmic existence also undoes the logic of the wondrous charm of dawn or the excited world beneath the still life. Arrhythmia is the Derridean supplement that undoes the logic of these charms and renders them all as nothing more than mere ornaments of modernity.

The failure of Marwood's hygienic quest is aptly summarised by the inclusion of the iconic movie poster from Harold Lloyd's *Safety Last* (1923) in the bathroom of the flat. The image depicts Lloyd holding onto the minute hand of a clock high above the city street. Thus, while insomnia begins as a light in the window above the street at 4 a.m. in Lefebvre's text, the film suggests a new residence for the insomniac. The image of *Safety Last* symbolically moves the insomniac, whose rhythms are defined by the 'fucked clock', into the liminal space of Lefebvre's balcony but without the structural support and perspectival vantage point that it offers. While sitting in the tub, Marwood offers his own translation of this anxious and precarious existence:

> Speed is like a dozen transatlantic flights without ever getting off the plane. Timechange. You lose, you gain. Makes no difference so long as you keep taking the pills. But sooner or later you've got to get out because it's crashing then all at once the frozen hours melt out through the nervous system and seep out the pores. (1987)

Marwood's arrhythmic endgame situates insomnia as a form of accumulation of fatigue and terror that results from his precarious grasp on the measured, rhythmic tick-tock of the clock. The use of substances (caffeine in the diner, drugs in the flat) is intended to manage this existence, or to extend it further, but these attempts at an alternative drug hygiene results in the further accumulation of fatigue and sleep loss in the body of the insomniac. It is a scene that eerily evokes Marx's notion that all that is solid melts into air.

Arrhythmia and pathology

In *Rhythmanalysis*, Lefebvre equates arrhythmia with pathology. While he refuses to attribute a cause to this state, he argues that it has its times and places and that it represents the dissolution of the regime of everyday life. Insofar as insomnia is strictly defined in the terms of an accumulation of sleep loss and the resultant schizophrenic mode of being, insomnia is thereby equatable not only to arrhythmia but to its definition as pathology. However, it is my contention that insomnia is not strictly equatable to pathology. In an earlier work, Lefebvre argues for a certain critical re-positioning of the philosopher who must be a 'witness to alienations, and their judge ... Keeping his [sic] vigil by night and day... [he will] not be satisfied simply to study the development of 'human nature'; he will want to help it, negatively at least ... by removing whatever may obstruct its fragile needs' (2008: 98). The notion of constant vigil establishes an equation between insomnia and the critical philosopher. The insomniac assumes this position through a re-reading of Lefebvre's balcony, in 'Seen from the Window,' via the metaphor of suspension over the city and the precarious grasp of time in *Withnail and I*.

Other philosophers have established an equation between insomnia and critical philosophy. In his interview, 'On the Utility of Insomnia', Emmanuel Levinas (2003) argues that insomnia is significant insofar as it represents awakening, which is counterposed to the negative attributes of sleep and falling asleep in Enlightenment thought. This strand of thinking dovetails with certain sleep scientists and a dominant mode of cultural thought, which sees sleep as a penalty for being awake. Sleep is no longer necessary since we have the technologies that allow us to extend diurnal rhythms, and since sleep is equatable to a lack of consciousness, perception and a sort of 'sleep of reason'. Awakening has thus been given a significant role and importance in modern thought, insofar as it represents the termination of sleep and the emergence of the light of awareness. However the phenomenological experience of insomnia is the antithesis of awakening. Insomnia is more aptly defined as the accumulation of wakefulness. Moreover, in insomnia there is no recourse to some sort of Levinasian 'search by the awakened for a new, more profound, philosophic sobriety' (2003: 127). Taking Marwood and Withnail as figurations of the twenty-four hour philosopher, it is obvious that they bear no resemblance to 'philosophic sobriety.' In contrast, Lefebvre's notion of the twenty-four hour philosopher sidesteps these problematic ascriptions to the experience of awakening.

As I have shown in regards to Marwood's narration beneath the image of Harold Lloyd, insomnia represents an accumulation of wakefulness that leaves the insomniac with not only a precarious grasp on time but also a physically felt accumulation of frozen hours within the body. Nathaniel Kleitman, in his experimental insomnia studies, examined the critical capacities of subjects who underwent periods of 40 to 115 consecutive hours of wakefulness. While he noted that pain, delusions and buzzing in the head accompanied these periods of accumulated fatigue and that pain figured as a cause of insomnia, he also noted that mental capacity remained the same, while attention waned (Kroker, 2007). Turning back to *Withnail and I*, it becomes apparent that what they are engaging in is a radically inattentive undoing of the logics of these spaces. Neither of them can sustain any activity or line of thought for long, but Withnail's denunciation of Regent's Park, and the modernist logic of leisure and restoration that it purportedly contains, is probably the clearest example of this critical capacity. Their critical vantage point takes its bearing from that 4 a.m. starting point and they bring this darkness with them into the heart of the modernist urban and rural topography. Thus, the accumulation of wakefulness also represents an extension of night into day, alongside the extension of light, as seen in Lefebvre's window, into the night.

In conclusion, what this insomnia-based re-writing of the critical philosopher entails is re-reading of Lefebvre's balcony. From the balcony, the attentive subject purportedly can listen and look at the rhythms below in order to 'dig beneath the surface, [to] listen attentively… [in order to see] the *objects* (which are in no way things) *polyrhythmically* (2004: 31). The essential starting point for Lefebvre is contained in this ability to grasp and restore rhythms, which represents a sort of Bergsonian synthesis of perceptive faculties, memory and heart. Marwood and Withnail do not occupy such a site. The solidification of wakefulness in the body entails an inability to pay attention or to form and recall memories. They also exist in a heightened state of terror insofar as their tenuous grasp on time is palpable, to the point that the heart that Lefebvre cites beats arrhythmically. However, they are not purely pathological. They bring forth critical approaches to the spaces and rhythms of everyday life. The experiences of Marwood and Withnail thus define an insomniac experience as an arrhythmic suspension over and simultaneous immersion within the street, the diner or the park. They also represent the social pathogenesis of schizophrenia, since the cumulative experience of this regime leaves them incapable of integrating to socially determined circadian rhythms of sleep and wakefulness. All that remains is the never-ending quest to immerse the insomniac self in the socially available rhythms, as Marwood articulates. Ultimately, Lefebvre's vigilant, critical philosopher has, at its core, an arrhythmic accumulation of wakefulness.

PART III
Mobile Rhythms

Chapter 8

'He Who Thinks, in Modern Traffic, is Lost': Automation and the Pedestrian Rhythms of Interwar London

Richard Hornsey

In August 1933, Dr F. Roberts of Cambridge published a letter in *The Times* disparaging the use of the word 'pedestrian' to refer to the ordinary walking person. There was, he acknowledged, no real alternative; 'walker' intimated strenuous hiking, whilst 'footer' and 'stroller' were plainly ridiculous. Yet as a more general common adjective, 'pedestrian' was already imbued with regrettable connotations:

> The last Budget, [he wrote], was dubbed in your columns 'a pedestrian budget.'
> If this implies that pedestrians are without imagination the metaphor may be in
> some degree justified. If it implies that pedestrians take no risks nothing could
> be farther from the truth. (11)

As Roberts' letter intimates, urban walking had become a hotly contested topic in interwar Britain. In the year of its publication, 597 pedestrians were killed whilst crossing the carriageway in the City and Metropolitan Police Area; evidence, it appeared, of the alarming mismatch between speedy modern motor vehicles and London's antiquated road network. With cars running alongside horse-drawn carts, electric trams and slow heavy lorries, the metropolis seemed beset by an increasingly entropic arrhythmia, manifest in black-spots of extreme congestion and an agitated confusion as diverse types of driver tried to negotiate each others' tempos and conditions of manoeuvring. In response to this urgent crisis, a coalition of administrative bodies, government departments, municipal bodies and interest groups tried to ameliorate London's traffic problems by orchestrating the movements of its various component units. Part of this involved a sustained and contentious supervisory attention to the motions of foot passengers and their perceptual modes of responding to the dynamic urban environment. Thus, if in its adjectival form 'pedestrian' implied a lack of imagination and an aversion to risk, then its ascendency as a noun during this period was entirely consistent. In what now appeared to many as the new 'motor age', pedestrianising the pedestrian had become a hegemonic project.

Henri Lefebvre foregrounds urban walking as a key example of 'dressage' or the breaking-in of the individual to (re)produce what he calls 'an automatism of repetitions' (2004: 40). Subjected to diffuse social discipline, the walking body orders itself via a regulated measure embedded in its musculature and exercised via a practised repetition that by-passes subjective volition. 'In the street', he writes, 'people can turn right or left, but their walk, the rhythm of their walking, their movements [*gestes*] do not change for all that' (40-1). The appearance of this rhythm as an innocent expression of the walker's unsullied physiology makes it harder, he argues, to recognise its profoundly historical and geographical origins – something now only recoverable by reflecting on the 'jauntier' movement of archaic pedestrians captured on archival film footage (38). Lefebvre roots this discipline in what he calls 'the military model' (39), a term that reveals his debt to Michel Foucault's earlier exploration of dressage in *Discipline and Punish*. Here, Foucault traced a mode of instrumental power developing within European armies from the seventeenth century; through a micromanagement of the soldier's body it became instilled with an 'automatism of habit' ([1975]/1991: 135), its accustomed gestures performing and policing the condition of its own subjection. This, Foucault argued, had hinged on the mechanism of the exercise, a carefully managed and repeated activity that invoked an optimal performance it never quite achieved. By enmeshing the body within a complex assemblage of instructions, supervisions, signals, architectures, classifications, time-tables and mechanical apparatuses, the military exercise served to correlate the soldier's physiological rhythms to the requirements of the wider system. It thus sought to produce an efficient and docile body, productively self-aligned through its own habituated measures.

Lefebvre is clearly aware of how contemporary pedestrians are precariously managed by a dynamic complex of both human and non-human actors. Gazing from the balcony of his Parisian apartment, he notes the rhythmic variance between those strolling tourists abruptly halted by solicitous street entertainers and the aggravated flows of peak-time pedestrians ordered by the automated pulse of traffic lights (2004: 29). Here, then, pedestrian rhythms are both systemically structured and contingently impacted by local particularities. Yet if the military manoeuvres of the eighteenth century provide an insight into the wider processes of modern discipline and its systematic correlation of corporeal rhythms, then the incorporation of urban pedestrians into this formation requires closer investigation. In industrial cities, this generally occurred sometime during the first half of the twentieth century as the purposes and meanings of the urban roadway were persistently, if unevenly, revised. Across the urbanised West, the street became culturally redefined no longer as a site of heterogeneous social activity, but as a functional conduit whose purpose was to facilitate the speedy flow of traffic. Historians have noted how the motorcar and its attendant ideologies of automobility impacted on the status and meaning of urban walking, whilst subjecting it to new cultural imperatives and forms of physical discipline (Bonham, 2006; Norton, 2007). Yet this chapter returns to the streets of interwar London to consider the mechanics of pedestrian dressage; for the importance of this moment lay not only in the reformulation of the highway, but in the attendant reorganisation of the walker's body and its rhythmic practices of movement and perception. Submitted to new modes of administrative ordering, pedestrians became re-imagined as modern urban subjects, corporeally invested with mechanisms of self-government and compliance that would productively bind them to the requirements of the corporate whole. In practice, this strategy encountered inexorable limitations, but its complex legacy still largely structures the ordered polyrhythmicity that constitutes the modern metropolitan street.

The periodical *Punch* had already intimated something similar within a cartoon published in May 1934 (Figure 8.1). Above the caption 'Why not a school for pedestrians?', Arthur Watts drew a large hall in which strapping instructors supervised a stream of pedestrians as they dashed chaotically between two rotating circuits of model cars. The suggestion here was that the accelerated tempo of passing motor vehicles had now entered into conflict with traditional ways of crossing the road. This task would have to be relearned; a matter of both speeding up one's movements and developing a sharper acuity concerning how and when to enter the carriageway. More significant, however, was Watts' casual invocation of contemporary innovations in motor manufacturing. His circuits of half-built cars, moving through the hall at their own steady speed, mimicked those moving assembly lines recently installed at such UK plants as Morris at Cowley and Ford at Dagenham (Turner, 1964; Burgess-Wise, 2001). First pioneered in 1913 at Ford's Highland Park factory in Michigan, the assembly line was the dominant motif within a major shift in industrial organisation, premised on a reconfigured relationship between human labour and the productive apparatus. Watt's crowd of middle-class urbanites forced to reorient their movements – with varying degrees of success – to the sudden demands of the automobile unwittingly parodied those Fordist factory workers simultaneously adapting their gestures to the very same thing. Connected to the motorcar at different moments in its lifecycle, the industrial labourer and the urban pedestrian were subjected to similar processes of enforced correlation. Both were located within new organisations of time, space, regulation and supervision that sought to realign their corporeal rhythms within a dominant assemblage of mechanical devices and informational flows. Yet the streets of London proved far less amenable to corporate management than the regimented space of the factory floor, which radically complicated contemporaneous attempts to automate the rhythms of pedestrian practice.

'When this man tells you to walk, you walk': scientific management and the automated body

To understand the connections between Fordist labour practices and the interwar mechanics of pedestrian dressage, I turn to the work of F.W. Taylor, an American mechanical engineer and leading advocate of industrial reorganisation at the turn of the twentieth century. His crowning polemic, *The Principles of Scientific*

WHY NOT A SCHOOL FOR PEDESTRIANS?

Figure 8.1   'Why not a school for pedestrians?', cartoon by Arthur Watts,
***Punch (Summer Number)*, 14 May 1934**

Source: Reproduced with permission of Punch Ltd. <www.punch.co.uk>

Management ([1911]/1998), still provides the most rigorous articulation of
modern industrial management and its rhythmic reorganisation of the active
human body. As an ardent reformer, Taylor's bugbear was what he termed the

'initiative and incentive' approach to industrial organisation, whereby workers learned their trade by observing more experienced colleagues on the job. Since labouring skills were largely 'handed down from man to man by word of mouth' and 'almost unconsciously learned through personal observation' (13), any task was invariably being done by a myriad of local 'rule-of-thumb' methods. More significantly, this placed technical knowledge solely within the body of the aggregate workforce, far outside the competencies of management. This, Taylor argued, was disastrous. Within an industrial structure based on maximising profits (and thus paying employees the lowest possible daily wage), the workers' interests lay in concealing from their employers just how productive they were capable of being. Thus manufacturing remained endemically dogged by 'soldiering', via both 'the natural instinct and tendency of men to take it easy' and the more 'systematic' deceleration of the entire workforce to the speed of their most tardy individual (6).

Against this, Taylor proposed his own system of 'scientific management', involving a major structural reorganisation of industrial production. Amongst the many 'rule-of-thumb' methods in operation, he reasoned, one alone must be the most soundly productive, and it was the employer's duty to determine and enforce this technique. This process involved three successive stages. First, the manufacturing process had to be disaggregated into its smallest component procedures, which Taylor christened 'tasks'. Then, deploying stopwatches and statistics, the most efficient way of performing each task was to be determined through the elimination of all wasteful corporeal or mechanical exertions. Finally, this optimum method had to be established as the absolute standard to which all workers must henceforth conform.

Once each component task had been properly ordained, management could deploy individual workers at different points in the production process as and when they were needed. As Taylor explained:

> The work of every workman is fully planned out by the management at least one day in advance, and each man receives in most cases complete written instructions, describing in detail the task which he is to accomplish, as well as the means to be used in doing the work. (17)

This marked an important inversion of the body's relationship to its productive activity. As the term 'rule-of-thumb' suggests, labouring practices had conventionally been guided by the contingent variability of workers' bodies, facilitating a more eurhythmic accordance between muscular exertion, perceptive responses and the tempos of machinery. Taylor's 'task', however, relocated this agency within a managerial abstraction. The labouring body now had to adapt to a prescriptive rhythm imposed from above, orchestrated via detailed instructions and arranged apparatuses, and which ultimately determined how much they were paid. Rest breaks too were to be carefully administered, not with reference to an individual's actual exhaustion but to the optimum rest breaks of that optimum labourer to whom they had now to aspire. As Taylor explained to Schmidt, a 'mentally sluggish' pig iron loader at the Bethlehem Steel Company:

> you will do exactly as this man tells you to-morrow, from morning till night …
> When this man tells you to walk, you walk; when he tells you to sit down, you
> sit down, and you don't talk back at him. (21)

Taylor's innovation was thus to bisect productive labour into two discrete activities. All elements involving some aspect of intelligence – deciding how best to arrange the machines, what speed to run them at, or when to take a break – were divested from the labourer and requisitioned by management. Even the most skilled machine operators, Taylor insisted, should not direct their own activities, for the time and effort this required would detract from their ability to achieve maximum productivity. Such calculations could only be done by a separate team of planners, preferably in an office away from the shopfloor, where they could focus unhindered on the task of planning each worker's schedule for the following day.

Every labourer's performance had to be monitored so that each could strive to match the required tempo, whilst managers could identify those underachievers too asynchronous to remain in employment. Yet, crucially, Taylor also recognised that such supervision would become less necessary over time as the worker's body become accustomed to its task and internalised its abstract set of optimum measures. Management thus became a matter of instilling the correct corporeal habits – the military exercise had colonised industrial production. In addition, by 1911 Taylor had already articulated the cybernetic structure of the modern corporation. Foreshadowing the digital computer that Alan Turing would map out a quarter of a century later (1937; 1951), all productive processes had been recast as an informational exchange between a commanding 'control' unit (the management) and an obedient 'executive' (the workforce). This self-regulating assemblage, sustained by its own internal flows of supervision and feedback, integrated each worker into an industrial apparatus that determined not only the goals to which their exertions were directed but the corporeal rhythms of their correlated labour. As Taylor proclaimed: 'In the past the man has been first; in the future the system must be first' (iv). By means of a series of abstract exercises, the labourer had been fully subjected via an automatic set of rhythmic gestures instilled within their perceptive apparatus and expressive musculature.

'No code of customs to guide them': the problems of pedestrian sovereignty

Taylor's writings did not directly influence the interwar drive to reform London's pedestrians, but did articulate the wider corporate logic through which its administrators came to approach the task of traffic management. Since at least the nineteenth century, the metropolitan highway had been culturally positioned as an emblematic space of national freedom; to move at will amidst its diverse populations was to celebrate a particularly English love of liberty and democracy.

Indeed, many Victorian modernisers had been significantly defeated in their attempts to improve traffic circulation by accusations of their assault on these principles (Winter, 1993). Early film reels seem remarkable today not only for the jauntiness with which London's walkers trot along its pavements, but for the apparent disorder of their uncoordinated trajectories. The largely horse-drawn traffic moves in two steady – if unregimented – flows, but pedestrians are seen entering the carriageway at will, to snake alongside and between its vehicles in what now appears as a precarious and incomprehensible anarchy.

During the 1920s, as the motor car gained in its ascendency, these archaic ways of walking came to register as fatally out of place. Two practices caused particular concern: individuals stepping into the roadway to avoid an obstacle on the pavement without due regard to passing vehicles; and substandard attempts to cross the road either through distraction, inattention or by misjudging the speed of oncoming traffic. Newspapers editorials, correspondence and the declarations of magistrates and coroners contributed to an ongoing debate in which blame alternated between selfish drivers out to deprive walkers of their rightful place on the King's Highway and those 'reckless' pedestrians who obstinately refused to adjust themselves to the requirements of modern traffic. Setting out to adjudicate between these positions, a loose conglomerate of public officials, administrative agencies and pressure groups sought to address what was now clearly recognised as the modern traffic 'problem'. Although the Metropolitan Police had been responsible for the everyday management of London's traffic since the mid-nineteenth century, the upkeep of its roads remained apportioned between numerous borough councils and local authorities. In 1924, however, the London and Home Countries Traffic Advisory Committee was created, gathering these bodies together with members of the Home Office, the Ministry of Transport and motoring organisations, to advise the Minister of Transport on traffic policy within the metropolitan area. This administrative rationalisation was recognised as the vital precondition for effectively managing London's roadways, a belief later consolidated by the dual promotions of Herbert Alker Tripp as Assistant Commissioner of the Metropolitan Police in 1932 and the forthright Leslie Hore-Belisha as Minister of Transport in 1934. Thus, by the mid-1930s, the improvised polyrhymicity of London's highways had become recognised as a legitimate object of bureaucratic management, something to be ordered and reorchestrated by a team of absent experts. By 1938, Tripp could confidently write of,

> a new science … which, as it takes shape, is found to be of a far-reaching character, embracing not only the immediate supervision of vehicles on the road, but also the problem of legislation, of public opinion and psychology, of road layout and equipment, of town and district planning, and many other matters. (1)

The metropolitan road system had been reconceived as a Taylorist factory writ large.

All this involved a foundational assumption about the functional purpose of the modern highway. As Tripp explained, traffic control had only two real objectives: 'to develop a rapid free flow of traffic'; and 'to prevent that rapid traffic from being an undue danger either to its own units or to the public at large' (1). The street was no longer an open stage on which a diverse populace improvised a picturesque democracy; it was, instead, a productive apparatus, still residually pluralist, but solely directed to the attainment of maximum safe speed.

Under the gaze of experts, the pedestrian had been quietly refigured as a variable component within a larger assemblage, their exertions to be managerially aligned with the requirements of the corporate whole. Traditional ways of negotiating the street were positioned as archaic and counter-productive, for old rule-of-thumb methods plainly invited catastrophe, congestion and death. As Mervyn O'Gorman, Vice-chairman of the Royal Automobile Club, remonstrated in 1929:

> Looking about I find one single traffic unit on the road whose movements are provided for by no legislation or custom, and for whom there is neither constant nor perfect control... That unit is the pedestrian, and this (apart from his frailty) is the chief cause of his sufferings. On the footway he wanders at will immersed in his thoughts, amusements, conversation. When he steps into the roadway he suddenly enters another world where the movement of every entity is, and must for safety be, controlled and foreseeable... walkers have no code of customs to guide them and to inform others as to their probable next movements. (13)

Prone to distraction and absent-mindedness, ordinary pedestrians had become sub-standard operators; they simply couldn't remain focussed on the job in hand. As a leader in *The Times* explained, car drivers were less prone to inattentiveness because driving remained 'a new and unusual' activity. Yet 'the walker is doing something which he learned to do in infancy and has been doing ever since,' which had penetrated over 'countless ages' to the depth of primordial instinct. According to this logic, motorists had already attuned themselves to the heightened perceptual responsiveness demanded by the car. Yet the pedestrian's body remained the site of an unstable conflict between habitual rhythms of (in)attention and the contingent irregularities of the mechanised roadway. What was needed, claimed *The Times*, was 'the penetration of the instinct by the sense of danger' (1927a: 13); an environmental responsiveness as automatic and unthinking as the reflex of putting one foot in front of the other. The paper continued: '(H)e who thinks, in modern traffic, is lost; until pedestrians act as 'unconsciously' as do all good drivers they must remain in danger' (1927b: 13).

These entreaties for pedestrians to develop a heightened 'road sense' would recur throughout the interwar period, for their 'hesitation' or 'faltering' remained an urgent and often fatal problem. On highways now dominated by the rapid but irregular tempos of passing motor vehicles, the pedestrian's traditional jurisdiction over when and where to step into the carriageway could no longer be endorsed. The lengthy distance covered by a car within its driver's normal reaction time turned such uncertain movements into a definite hazard. Hence the idiosyncratic rhythms of individuals on foot had to be rendered entirely predictable.

Throughout the 1920s and 30s, London's walkers were subjected to successive attempts to choreograph the rhythms of where, when, and how they moved through their city's streets. Common to all was the disaggregation of walking into a component set of separate tasks. Proceeding along the pavement and crossing the road became conceptually distinguished, opposed to each other and discretely managed within contiguous and controllable sequences. This involved a complex ensemble of mechanical devices, official personnel and regulatory imperatives, all of which sought to produce a newly automated pedestrian subject fully correlated with the ordered polyrhythmicity of managed modern motor traffic. However, in practice, the contingent rhythms of less controllable elements within the metropolitan highway persistently intervened, effectively frustrating its more simplistic imaginings of repetition and control.

'Cross Now': the technologies of pedestrian automation

The pragmatic limitations of pedestrian discipline became evident very early on. In 1917, the newly-formed London 'Safety First' Council began an earnest campaign to retrain walkers to 'Keep to the Left' of the pavement. If this measure was universally adopted, they argued, then those nearest the kerb would always be facing the oncoming traffic and thus less likely to step into its path. By 1918, the council claimed the support of 24 out of London's 29 local authorities, some even erecting signs to instruct pedestrians accordingly. Yet despite the irrefutable logic of this proposal, it consistently failed to persuade either the government or the police. As one Metropolitan Police memo of 1919 explained, the automatic behaviour it hoped to instill faced insurmountable difficulties:

> ['Keep to the left'] is the reversal of an instinct inherited for generations ... Instinctively we move to the right when meeting and we are so constituted physically that if lost in a fog or in darkness we shall almost certainly move to the right.

Like the more general problem of acclimatising pedestrians to the demands of the motorised highway, the walking body was understood as already endowed with its own primordial rhythms. Resynchronising these would require relentless supervision, involving considerable manpower and political resolve. In the absence of these things, a comprehensive programme of pedestrian dressage was already unachievable.

Thus traffic managers sought alternative means to prevent pedestrians from drifting in front of vehicles. In 1930, the Traffic Advisory Committee recommended the painting of a continuous white line parallel to the kerb to remind walkers of the immanent danger. More selective in its supervisory address, this measure was similarly opposed by the police on the grounds that adaptation to the line was likely to occur far more rapidly than adaptation to the perils it announced. Indeed, Tripp was defiant that the only real solution lay in the total physical segregation of pedestrians and vehicles. Under his influence, a mile of metal railings was opened in May 1936 on both sides of East India Dock Road in Poplar, the first section of a scheme that would eventually stretch for three continuous miles. As Hore-Belisha explained to the press, such guard rails would 'by physical necessity create a practice which seemed psychologically difficult' (*The Times*, 1936: 11). Thus, if the residual rhythms of pedestrian (in)attention were too engrained to be comprehensively resynchronised, such railings promoted a more forceful disciplining of walking practice. As a mechanical apparatus, they effected a permanent and impersonal supervision that divested the pedestrian of any residual self-government. Yet they were also very expensive and – away from the marginalised districts of the impoverished East End – proved contentious to implement. Despite Tripp's emphatic advocacy, their use remained selective.

Efforts to automate pedestrian behaviour thus came to concentrate on the segregated task of crossing the street. The central innovation here was the designated crossing place, a complex assemblage that sought to order where individuals entered the road and thus cleanse the rest of the carriageway of errant pedestrian activity. The first concerted experiments began in December 1926, when signs bearing the words 'Please Cross Here' were erected across the tributaries to Parliament Square. This had recently been converted into a 'one-way' gyratory system, in which – foreshadowing Watts' cartoon – the accelerated circulation of clockwise vehicles had made crossing the road a great deal more hazardous. Between the signs, a white line was painted across the carriageway to inform drivers where to stop when so directed by the police, thus producing a 'safety lane' across which the free flow of vehicles was alternated in a binary rhythm with the passage of unhindered pedestrians. On 11 June 1934, over a hundred such crossing places were launched at junctions around Westminster, Camden Town and Mile End, all under the jurisdiction of either pointsmen or automated signals – the first phase of what was envisaged as a holistic system that would eventually cover the entire metropolis. Ultimately, Hore-Belisha hoped to secure legislation that would prohibit pedestrians from entering the carriageway other than at such designated crossings. Fearing that any distance of more than 200 yards between them would spawn legitimate complaints about excessive deviation, he also pushed through the creation of additional 'uncontrolled' crossing places, governed not by police officers or automated signals, but solely by a regulatory code of conduct. These were marked on both pavements by a metal pole surmounted by a yellow glass globe, which – with shameless self-publicity – the Minister immediately christened 'Belisha beacons'.

The most pressing dilemma here concerned how to ensure that pedestrians acknowledged the authority of these devices and restricted themselves to the designated safety lanes. The initial proposal focussed on erecting signs at a distance from the crossing to mark out a 'sterilised' space within which entering the road would now be forbidden. Yet police concern about their lack of enforceability led to an ongoing experiment with alternative methods of supervision. From December 1936, special officers routinely distributed printed paper slips to those individuals seen wilfully trying to traverse the street in the vicinity of a crossing. Mimicking Taylor's scheme of written instructions, these informed pedestrians of their substandard behaviour, but crucially avoided confrontation by replacing direct orders with a more generalised plea for responsible behaviour. Similarly, three years later, when so-called 'courtesy cops' were stationed at several key crossings armed with portable amplifiers and loudspeakers, officers announced only a generic recommendation to use the safety lanes rather than to chastise specific individuals (McKenzie, 1939). Outmoded notions of pedestrian sovereignty thus set clear ideological limits on how explicitly London's pedestrians could be managed and Hore-Belisha's foreshadowed legislation never came to fruition. Indeed, metal railings became the apparatus of choice for sterilising the space around designated crossings, their tacit and impersonal injunctions marshalling walkers towards the lanes with less fear of retribution.

The pedestrian crossing place contained a clear imperative towards automating the responsive behaviour of walkers – wait until given the directed signal, then proceed across the carriageway – but the apparatus itself was perpetually frustrated by both the irregularity of London's traffic flows and the archaic layout of its road network. One residual problem concerned how exactly the pedestrian should interpret the orders given. A frequent early complaint was that pointsmen were found to habitually release traffic before individuals had reached the other side. Yet as Tripp made clear, officers could not wait for every pedestrian to mount the opposite pavement for, on busy crossings, a heavy flow of walkers would prevent vehicular traffic from progressing at all. Officers were thus ordered only to signal to pedestrians their intention to release stationary vehicles; it was the responsibility of the pedestrian to monitor these signals and calculate for themselves whether they had sufficient time to cross.

Likewise, there was an ineluctable ambiguity concerning how London's pedestrians should respond to automatic traffic lights. Throughout the 1930s, decisions were routinely made not to add pedestrian-facing signals to such devices, for it was generally recognised that walkers would be unlikely to stand waiting for authorisation if a sizable gap appeared in the traffic. A persistent contravention of the 'Don't Cross' command, it was feared, would jeopardise the authority of the entire assemblage. Thus official regulations reluctantly conceded that pedestrians were at liberty to traverse crossings at any time, providing they did not 'hinder the free passage' of any oncoming vehicles (Minister of Transport, 1934). In addition, whilst such regulations stated that those vehicles turning into a tributary sidestreet from the line of proceeding traffic had to give way to any crossing pedestrians, it was overwhelmingly felt that a direct order to 'Cross Now' would foster in walkers a misplaced and dangerous sense of invulnerability.

Time and again, the responsive automation built into the design of the
controlled crossing place was undermined by the unavoidable contingency of the
environments in which they were erected. To this end, experiments were made
in 1938 with a number of 'compulsory crossings' that forced pedestrians to wait
behind metal barriers until the controlling officer signalled for a colleague to
raise them. That observers should note a general acquiescence with the devices
suggests that, by the late-1930s, pedestrians were becoming accustomed to such
temporal management (*Evening News*, 1938). Yet such compulsory crossings were
soon dropped on account of their heavy running costs. Despite the endeavours
of London's administrative authorities, then, the dangerous unpredictability of
its motorised highways demanded the retention of some element of pedestrian
initiative, a more complex set of environmental responses that exceeded automatic
reflexes.

Similar difficulties were encountered at uncontrolled crossings away from the
direct jurisdiction of pointsmen or traffic lights. Here pedestrians had a clear right
of way, since regulations required 'the driver of every vehicle ... to proceed in
such manner and if necessary to stop so as to allow free passage to every pedestrian
who is crossing the carriageway' (Minster of Transport, 1934: 2). Yet uncertainty
persisted over how both parties should interpret each others' movements, for
there were no codified signals for expressing a desire to cross – aside from one's
proximity to the crossing point itself. In July 1937, an Appeal Court earned the
wrath of motoring organisations by ruling that since pedestrians had legal priority
at uncontrolled crossings, they could never be found negligent even if they stepped
into the direct path of an oncoming vehicle (McKenzie, 1937). Commentators
agreed that such forthright adherence to the regulations invited catastrophe, and
called for all parties to exhibit a greater 'give-and-take' when negotiating the
crossing – an informal mutual courtesy that resisted codification as either a set of
rules or mechanistic signals.

The temporal irregularities of London traffic caused further complications, for
at certain uncontrolled crossings a continuous stream of pedestrians risked holding
up vehicles indefinitely. Thus at selected points, Belisha beacons were eschewed in
favour of pedestrian-activated push-button signals, which halted the traffic for a clear
20 seconds, followed by a compensatory two minutes of uninterrupted vehicular
passage (*Evening Standard*, 1931). These devices formalised mechanically those
semi-improvised alternations between pedestrians and drivers, but uncertainty still
persisted. The two-minute buffer period encouraged more impatient pedestrians
to cross prematurely against the signal, such that when the lights did change,
vehicles were frequently halted when there was no-one waiting to cross. Indeed,
additional sensory pads were often embedded into the surface of the carriageway
precisely to detect, on each and every occasion, whether the two-minute delay was
actually required. Thus, the rhythmic variability of London's vehicular and foot
traffic undermined the more straightforward attempts to choreograph its activities
according to an ordered binary measure. Only by means of a highly complex

assemblage, sensitive to a range of contingent factors, could a system of direct orders become operationally viable.

Conclusion: from the automatic to the cybernetic pedestrian

The complexity of London's road network and its rhythmic irregularities thus worked to frustrate attempts to automate pedestrians and recodify their practices as a disciplined set of habituated responses. Any instruction to 'Cross Now' remained dangerously direct, whilst uncontrolled crossings had always to be supplemented by an informal mutual courtesy. Those devices that worked most directly to discipline the walking body – guard rails, 'compulsory crossings', and pedestrian-activated vehicle-actuated traffic signals – were only selectively deployed due to their prohibitive cost. Thus, within the dynamic environment of the modern highway, ordinary walkers were positioned as more than unthinking automatons – something which most regulations and apparatuses acknowledged. By the close of the 1930s, pedestrians may have become accustomed to using designated crossing places, but they retained a formal and practised autonomy over the places and times at which they entered the carriageway.

London's interwar pedestrians were thus refashioned as strangely hybrid creatures. Whilst more generally aligned to wider corporate objectives, the contingencies of the metropolitan street demanded a more sophisticated form of predictable behaviour. Pedestrian management became less about habituating certain corporeal actions, than about instilling a type of intelligent responsiveness that could readily adapt to all unforeseen circumstance. This moved the underlying model of British pedestrian management beyond the docility of the Taylorised labourer, towards the more complex intelligence of the cybernetic subject that Norbert Wiener would articulate at the end of the Second World War (Wiener, 1950; Galison, 1994). Encapsulated most succinctly by the figure of the anti-aircraft gunner, this approach conceived of the individual not as a servile labourer, whose corporeal rhythms were bound to the regular measure of the commanding machine, but as an expressive element within a wider assemblage, in which the human subject and the mechanical apparatus worked in conjunction to ameliorate the impact of unpredictable variables. Whilst geared towards an outcome that was fully aligned with systemic objectives – safely crossing the road in a manner that didn't upset the smooth passage of motor traffic – each individual street-crossing remained a complex and variable performance, as pedestrians engaged with those other devices to achieve a stable (though never fully predictable) eurythmia with the highway. This residual element of individual sovereignty came to define the London pedestrian in the twentieth century, perhaps more so than in comparable cities in the US and Australia (Norton, 2007; Bonham, 2006). Exceeding the docility of disciplinary dressage, the perpetual and ongoing synchronisation of the pedestrian into the contingent polyrhythmicity of London's streets pointed to a newly cybernetic form of urban engagement, an ordered but ultimately less determined type of responsively programmed 'road sense'.

Note

The unpublished sources used in this chapter reside at the National Archives, Kew (TNA/), within the following files: MEPO/2/4666; MEPO/2/4730; MEPO/2/4715; MEPO/2/4724; MEPO/2/7210; MEPO/2/7374; MEPO/2/8035; MT/128/67; MT/128/67; MT/34/223; MT/34/253; MT/34/254. This archival research was made possible by a generous Small Research Grant from the Royal Geographical Society (with Institute of British Geographers).

Chapter 9

Improvising Rhythms: Re-reading Urban Time and Space through Everyday Practices of Cycling

Justin Spinney

Introduction

> If you look at *The Highway Code*, there's about three of the regulations which actually refer to cycles. It was not designed for cyclists, it was designed for car drivers; the whole traffic light system, everything is designed for car drivers and perhaps for pedestrians, and cyclists aren't part of the way that the system is designed and so you have to make your own way. (Noam, interview: 21 September 2005)

The opening scenes of the film *28 Days Later* (2002) begin with an aerial shot looking North over Waterloo Bridge and the Embankment in London. Shot in the early daylight hours the scene is devoid of morning commuters, traffic and pedestrians; there is no movement whatsoever. The complex rhythms which define the modern western city are entirely absent in this scene and the feeling conveyed is unsettling and alien. The place depicted appears unfamiliar and strange not because its spatial arrangement has changed, but because its rhythmic arrangement has changed: the scene makes mundane rhythms visible through their absence.

Urban 'commuting' is one example of an everyday practice rendered virtually invisible because of its familiarity. It is conceived by those who design infrastructure as a utilitarian practice concerned with getting from A to B as quickly as possible. As Edensor (2009: 4) notes, 'commuting involves the channelling ... and scheduling of a privileged form of mobility within managed "routescapes"'. The spaces, timings and hence rhythms of the road network thus reflect a bias towards functionalism but also towards motorised vehicles. When encountered using a different 'instrument' of mobility – the bicycle – the appropriateness of many of the rhythms embedded in the built environment to facilitate the rapid passage of car-drivers are called into question. Consequently many cyclists employ different rhythms; weaving time and space together in unintended and 'inappropriate' ways in order to navigate the city and produce specific spatio-temporal orderings.

Conceptualisations of movement and mobility in geography are increasingly emphasising rhythm as a way of investigating the interplay between temporal and spatial dimensions. Through an exploration of urban cycling practices in London, this chapter contributes to this agenda by focusing on the ways in which the dominant rhythmic orderings embedded in the design of public and road spaces are manipulated and subverted through everyday practice. This is not to say that all cyclists improvise rhythms in the same way, far from it: cyclists are a heterogeneous bunch and their actions are influenced by skills, dispositions and other cultural factors. Thus whilst the practices of some cyclists suggest that particular flows and rhythms extend from a playful and highly skilled interaction, the actions of others suggest that a stop-start staccato rhythm may stem from a more nervous and discursively entrained encounter. Whilst acknowledging this variety, here I concentrate largely on the role of embodied experiences in affording the rhythmic variation of hybridised subjects.

I begin by briefly discussing how the rhythms that govern the flow of mobilities within the city are premised upon the affordances of particular 'instruments' – motorised vehicles, yet within the modern transport system, road users employ a number of different instruments to play the same 'piece'. I argue here however, that cyclists (as one example) are expected to perform largely in the same way as their motorised counterparts despite experiencing strikingly different affordances and possessing divergent capabilities. Accordingly, many cyclists fail to adhere to the dominant rhythms prescribed by system designers; rather they improvise rhythms which often contravene what is written and appropriate. Rather than theorise these re-readings of urban time-space as resistant or criminal, I suggest that these improvisations emerge because the hybrid bike-rider is oriented to the world differently from the car-driver. In the second section of the chapter therefore, using examples drawn from empirical fieldwork with cyclists in London between 2004 and 2006, I identify three 'immaterial' experiences – duration, energy and vulnerability – which account for the improvisation of rhythms by urban cyclists.

Laying down the rhythm

According to Sennett (1994), the 'healthy' circulation of the city has long been ordered according to the instrumental rationalities of economy; its spaces and movements ordered scientifically and functionally to facilitate efficient production and consumption. The modernist city is thus often theorised as colonial, where technologies of planning and architecture come together to build new societies and indoctrinate citizens within the spatial confines of rationally planned towns. The movements of citizens are thus underpinned by the logics of functionalism whereby the, 'forms of buildings should reveal their structural roles ... and instill moral and ethical ideals in those who see and use them (Hill, 2003: 10). For Lefebvre,

> space lays down the law because it implies a certain order – and hence also a certain disorder (just as what may be seen defines what is obscene) [...] Space commands bodies, prescribing or proscribing gestures, routes and distances to be covered. (1991: 143)

However, Yelanjian (1991: 91) writes that power is not only dependent on the division of space but also upon the ordering of time to create a time-space. Geographers and sociologists have long been concerned with the interweaving of time and space; the work of Hagerstrand (1977) in particular has added a new dimension to static and two-dimensional readings of the city. However, whilst Hagerstrand had much to say about everyday life and practice, Mels (2004: 16) notes that his conceptions contrast with humanistic geography's concern with the phenomenological, meaning and subjective experience.

Similarly, Marxian geographers (Harvey, 1989; Jameson, 1991) have argued that contemporary urban rhythms are distinctively capitalist rhythms. Harvey for example has suggested that new systems of transport have disrupted temporal and spatial rhythms. However, in common with Hagerstrand's time geography, Marxian readings of time-space have rarely articulated heterogeneous temporal experiences.

In more humanistic accounts, part of the character of place lies in the temporal structure of space, or 'time-space', whereby 'timed-space is the essence of place; that it is the timing component which gives structure to space and thus evokes the notion of place' (Parkes and Thrift, 1978: 119). Building upon this understanding Buttimer (1976: 289) asserts that synchronisation of time and space can grasp the dynamism of the lifeworld.

More recently, Lefebvre has called for time and space to be thought together in the form of rhythm. For Lefebvre, rhythm unites elements which 'mark time and distinguish moments in it' (2004: 9). Whilst he outlines a number of forms rhythm can take, of particular note are secret or internal physiological rhythms and external or 'dominating' rhythms which aim to affect bodies and materialities beyond themselves (17-18). As Lefebvre (1991) notes, rhythmic animation should not be confused as purely mechanical or as socially structured in a deterministic manner, rather it is the product of reciprocal interaction and (often) mutual constitution, between mechanical and 'natural' rhythms.

Given that rhythm implies the association of a space and a time and thus movement, there is surprisingly little mention of it in mobilities research. Until recently, what little geographical work there has been on rhythm or time-space has largely been the preserve of transport geographers. Whilst such work has provided valuable insights, it is generally concerned with macro-scale orderings of everyday mobility which attempt to explain the number of trips people make on a given day and variations in any rhythmic pattern which emerges (Axhausen et al, 2000; Novak and Sykora, 2007). Such accounts are geared toward understanding the relationship between urban structure, the transportation system and household travel patterns (Timmermans et al, 2003) and little attention is given to the way that the rhythms of the journey itself are structured (or contested) through discourse, experience or formed through the situated knowledges of hybridised agents. Rather the emphasis in these accounts is on explaining human movements and rhythms with reference to material and instrumental factors (for a critique see Spinney, 2009).

Moving beyond these somewhat limited explanatory accounts, there has recently been renewed discussion of the interplay between the material and immaterial in cultural geography (Anderson and Tolia Kelly, 2004; Latham and McCormack, 2004; Whatmore, 2006). Certainly as Crouch has noted, one of the key contributions of the cultural turn has been to disclose a range of 'new materials' based on the assumption that 'meaning is produced in the encounter between human subject and place, and other human subjects and a range of material artefacts' (2000: 73). Crouch's point suggests the importance of considering the interplay of bodies, technologies, environments and the affective and sensory nature of human subjects in the construction of meaning. For Lefebvre (2004) this interaction of the material and the immaterial lies at the heart of 'rhythmanalysis'. Accordingly, geographers are increasingly paying attention to rhythms in space (Bissell, 2007; Crang, 2001; Cronin, 2006; Edensor and Holloway, 2008; Edensor, 2009; Lorimer, 2007; May and Thrift, 2001; McCormack, 2002; Mels, 2004).

Moving the beat

As this emphasis on rhythm implies, it is the interweaving of time and space in the form of various material and socio-legal signals that lays down commands in the built environment, defining how we should move around, setting out desired forms of conduct and framing what movement means by making certain movements easier to perform than others. Certainly as Markus argues, architectural design standardises routes, shaping pathways which direct and discipline movement (cited in Kofman and Lebas, 1995). Boyle elaborates, noting that numerous technologies discipline and direct flow in the city such as street signs, pavements, islands, traffic lights and road markings. These 'boundaries are mechanisms that frame, limit, and define their subject by immobilising relations of difference in order to freeze meaning' (2004: 5); regimes of discipline that 'work upon urban circulation by constructing and coding authorised lines along which actual mobilities are to be channelled' (8). Such standardisations are therefore more than just discursive constructs; they are embedded in many features of the built environment and act to enclose, confine and organise behaviour in particular spaces. Murdoch (1998) defines such standardisation as a spatial, temporal or spatio-temporal segmentation of the world.

Within this spatio-temporal ordering, not all rhythms are deemed equally desirable. An infrastructure has been developed which seeks to segregate different styles of movement and to ensure that movement should cease at certain intersections to minimise the possibility of collisions, laying down a particular rhythm as 'appropriate' and criminalising others. Street architecture therefore can be theorised as the reification of particular values, and whilst these crystallise out of process, they simultaneously take on an 'objectness' at the point of crystallisation which lends them a permanence. Designed to embody particular actions, material 'things' lend themselves to being practised in a certain way; facilitating some actions whilst making others harder.

However, this permanence is not given; rather it is lent by the thing being repetitively practised in a certain way. Actors also take their cues from those around them as knowledgeable about the correct modes of conduct. The result is that particular actions pass 'as natural precisely when [they] conform perfectly and without apparent effort to accepted models, to the habits valorised by a tradition' (Lefebvre, 2004: 38-9). As Edensor and Holloway point out, the weaving together of numerous rhythms produces distinct forms of spatio-temporal consistency: the diffuse working of power seeks rhythmic conformity through the ordering of normalised interpretations about when particular practices should take place (2008). Accordingly institutional rhythms continually emerge through repetitive practice.

The incentive to engage in these desired practices is further backed up by discursive and socio-legal framings such as *The Highway Code*. In common with other Taylorist standardisations of movement, such framings are geared towards 'the transformation of learned habit – embodied habitual mobility – into a rigorous and scientifically coded abstraction of human motion' (Cresswell, 2006: 86). For Imrie and Thomas (1997: 1402), such readings uphold the notion that the law is 'beyond reproach in occupying a higher, sanctified, playing field where it operates to ensure just and equitable outcomes (for all involved)'. 'Appropriate' rhythms as legally framed therefore fail to adequately account for the importance of context and embodiment in defining rhythm, assuming instead that such specific rhythms are equally available to all.

Specifically then, material 'carchitecture' naturalises particular rhythms through a combination of objectness, discursive framing and habitual practice, facilitating some actions whilst weakening others to produce 'appropriate', predictable, normative flows. Mobility therefore often produces a rhythmic repetition through practised knowledges premised upon what is expected and appropriate within particular contexts. Thus as Yelanjian notes, the power produced through the ordering of a time-space is not solely embodied in individuals or systems, rather it is produced as a dispersed ordering force (1991). The result according to Lefebvre (2004) is that the rules of traffic control constitute a rhythm which the mobile subject is expected to follow.

Accordingly, processes of ordering and standardisation attempt to organise a system based upon the affordances of motorised vehicles, in which the architecture of the built environment has increasingly assumed and materialised a normative range of instruments (implying a method of propulsion and level of vulnerability) commensurate with motorised transport. The results have been that the rhythms of the modern city are overwhelmingly premised upon the body and affordances of the hybrid car-driver; other users such as cyclists and pedestrians have increasingly been expected to conform to the rhythms laid down for the benefit of motorised transport (Bendixson 1974; Packer, 2008; Urry 2000).

Improvising rhythms

The corollary of this rhythmic closure is that alternative rhythms become harder to imagine. Certainly as Amin and Thrift (2002) note, the city is rarely subject to rhythmic chaos because rhythms become collectively and individually repeated in order to negotiate urban life. However some users, including many cyclists, regularly ride off the 'score'; going when they should stop, weaving when they should be stationary, and using spaces 'inappropriately'. Whilst patterns or spatial structures are the outcome of long-term development and their basic character is reproduced and confirmed by everyday human activities, they are also reshaped by those human activities that do not correspond to existing patterns (Novak and Sykora, 2007; Cronin, 2006). In any space, as Edensor and Holloway go on to note, the rhythms of any number of different actors intersect to give a place its character. Certainly the 'apparently repetitive and regular rhythmic patterns are apt to be punctured, disrupted or curtailed by moments and periods of arrhythmia' (2008: 485).

Many accounts position any arrhythmic improvisation as a form of active resistance (Borden, 2001; de Certeau, 2002; Flusty, 2000). Edensor and Holloway (2008) rightly flag Lefebvre's tendency to frame the duality of structure and agency as inherently resistant, for while he acknowledges polyrhythmia, Lefebvre spends much of his time critiquing official and capitalist rhythms and the ways in which these rational and calculative rhythms impose themselves over the rhythms of the body. My intention is not to deny that many 'inappropriate' rhythmic cycling performances are intended to be resistant; one only has to look at the Critical Mass monthly bike rides around the world to see evidence of intentionally resistant rhythms. It is rather to suggest that there are other less sensational and oppositional ways of understanding such practices.

May and Thrift, for example, argue that time appears differently according to particular spatial arrangements within a given setting, suggesting that 'a sense of time emerges from our relationships with a variety of instruments and devices' (2001: 4). The way in which time-spaces are produced and experienced is thus a product of how we are oriented to the world and as a result time-spaces and rhythms are constituted through processes of embodiment and hybridity. Moreover, matter is constituted through the rhythm and routine of practical actions (Anderson and Tolia-Kelly, 2004). May and Thrift (2001) note that accounts have generally failed to look at the different perspectives generated through the way gendered, aged and technologised bodies experience time-space. Technology is foregrounded here because it becomes the instrument through which the score is interpreted and mediated. Edensor and Holloway flag its importance commenting that, 'mobile experience reconfigures the landscape in accordance with the technology of travel and the affordances of space' (2008: 488; see also Ingold, 2000). Wajcman (2008) further points to the creation of new social relations through technologies, suggesting that rather than simply 'saving time', new technologies change the nature and meaning of tasks, creating new material and cultural practices. Indeed, the role of different technologies in mediating experiences of time is well documented (Cresswell, 2006; Massey, 1992, 1993; May and Thrift, 2001; Thrift, 2004; Urry, 2002, 2003).

In Edensor and Holloway's account of coach touring, 'the technologies of the coach are trajectories of ordering that can similarly configure particular ways of experiencing space and produce distinctive rhythmic patterns' (2008: 488). The technologically hybridised body is thus of central concern when considering the way in which rhythms and arrhythmia are produced and contested because it orients us to the world in a certain way, opening up affective and sensory possibilities which influence the way in which we negotiate space and spend time.

Crang (2000) uses rhythm to evoke the notion of an urban music in order to rethink architecture and infrastructure as more than just frozen time. The association with music is quite apt because it is analogous to the ways in which 'scores' are created which set out appropriate rhythms and instruments for particular performances such as commuting. However, it also allows an exploration of what happens when that score is played using a different instrument from that which was intended. The resulting improvisations, syncopations and arrhythmias help to highlight which instrument the score was intended for and why it might be played differently using another instrument.

I argue that one of the key reasons why the rhythms laid down by many cyclists do not conform with the official and desired rhythms laid down for motorised vehicles is not because they resist the official ordering of space and time. Rather it is because the hybrid bike-rider is oriented differently to the material and immaterial affordances of the urban environment and thus many cyclists are engaged in a process of making do which involves improvising rhythms in order to navigate the city.

Rhythmanalysis thus entails paying attention to the body and the senses and moreover, to the instrument being employed to interpret the score and the rhythmic variations that this can produce. I discuss here how the rhythms of cycling are structured through different experiences of time, vulnerability and energy. Understanding these is vital to explaining 'inappropriate' rhythms as other than a simplistic rejection of dominant rhythms. Rather these improvised rhythms are the product of hybrid bodies that both possess and perceive affordances other than those embodied and facilitated within the dominant ordering of the city. Whilst numerous accounts of time-space compression suggest that modern forms of movement have radically altered the experience of time-space for all (May and Thrift, 2001), as I demonstrate here, the accounts of cyclists suggest that there is still a diverse range of rhythms produced and experienced simultaneously.

Timed-space: duration

> ... if I wanted to be stuck in traffic I'd drive to work. (Alan, 21 October 2004)

In the search to minimise serious collisions and keep traffic flowing, time conceptualised as a quantifiable variable is often overlooked by planners and engineers with little thought for different experiences of time or the different options open to people on how to 'spend' waiting time. One of the consequences of this ordering is that it forces the car-driver into a spatially linear, stop-start rhythm at signalised junctions. Whilst queuing in such instances, drivers can engage in various practices to pass the time – listening to the stereo, making a phone call, doing their make up or talking to a passenger. If they have practical knowledge of an area they may also be able to keep moving by finding an alternative route (Hagman, 2006). The bike-rider however may perceive the affordances of the situation very differently, exhibiting different ways of dealing with time and thereby constructing very different time-spaces.

As Cresswell (2006: 5) notes, 'mobility as a social product, does not exist in an abstract world of absolute time and space, but is a meaningful world of social space and social time'. Accordingly, different forms of movement and mobility produce particular temporal experiences as they orient the subject differently in relation to the space of time. Certainly Crang (2001) identifies a difference between lived and represented time, and Watts and Urry (2008) note the disjuncture between perceived travel time and lived travel time, contending that travel time is not the counting of minutes but is embedded in tasks. To say something takes 10 seconds for example, is wholly different to the potential experiences of those 10 seconds. To illustrate this point Henri Bergson used the example of watching sugar dissolve in water, noting that waiting for the sugar to dissolve was not simply the mathematical time that it takes to do something that can be universally applied; instead, 'it coincides with an impatience that constitutes a portion of my duration and which I cannot protract or contract at will' (cited in Pearson, 2002: 10). Thus time is always perceived subjectively and differentially rather than absolutely, and every individual has a personal sense of time (Carlstein et al, 1978a, 1978b). Bergson also claimed a distinction between the objectification of time through vision as a series of snapshots or transitions, and the embodied experience of time as a felt and unquantified continuum or 'duration'. He suggested that this was because,

> as muscular sensation, they are a part of the stream of our conscious life, they endure; as visual perception, they describe a trajectory, they claim a space [...]. The latter is divisible and measureable because it is space. The other is duration. (1999: 35)

Thus for Bergson embodied movement is central to the experience of time. With this understanding in mind, I contend that there are differential modes of waiting exhibited by cyclists, but the one I want to focus on here is that of *filtering*. In the same way that a knowledgeable driver may avoid a traffic queue by taking a different and often circuitous route, the cyclist often achieves this at a microscale, avoiding having to stop by taking a circuitous route through a traffic queue. According to de Certeau (2002: 18), there is a 'tactical and joyful dexterity' associated with the mastery of such a tactic which enhances the pleasure of 'getting around the rules' of a constraining space. By way of example, in his journeys to and from work, Alan frequently used filtering as a time-space tactic, constantly weaving in and out of stationary and slow moving traffic.

Figure 9.1   'Passing time in the traffic' (author's collection)

> I can't put it (weaving) down to anything really [...]. I think it's the fact that if you're stuck behind traffic you're sitting there basically sucking their exhaust, so I might as well be going through the traffic rather than just sitting there lapping it up. [...]. I hadn't realised I did it as much [...]. I think it's because I'm not going the speed that I want to, I'm restricted by what they're doing even though they're moving forward. Being stuck in the traffic is really boring too; if I wanted to be stuck in traffic I'd drive to work. (interview, 21 October 2004)

Alan's account suggests a situated understanding of what the bike-rider might reasonably be capable. Alan uses the spaces between vehicles to keep time 'flowing': whilst the rhythm conceived and laid down by the system's designers in such places produces bursts of rapid movement punctuated by rests for car-drivers, his overlain rhythm is slower and less spatially linear. Filtering, although slower than freely cycling down a clear carriageway and hardly quicker overall than stopping and pulling away, is seen by Alan as preferable to stopping because the duration experienced in moving through space is favoured over the experience eof being stationary. Lefebvre offers the example of appropriated time to suggest a duration where 'time no longer counts' because the subject is immersed in activity of one form or another (2004: 76).

It should be noted that as a tactic, filtering is not equally open to all because it involves a degree of balance and bike control, and if performed at speed may entail a greater risk to the rider than if they were static. Coupled with a relative absence of formal disciplining, Alan's confidence and skill with the bike highlights the construction of alternative time-spaces and rhythms less conditioned by the dominating rhythms of road space. As Thrift (2004) notes, myriad adjustments and improvisations open new lines of flight, producing the fabric of time-space as open-ended.

Bodily rhythms: energy

> the *main* limitation of (the contemporary transport planning) approach stems from the assumption by its installers that people will behave like obedient automatons. A brief period of observation of a junction will confirm that they do not. Pedestrians routinely disregard red lights (in Britain red lights for pedestrians are merely advisory, not mandatory). And they are frequently found on the wrong side of the barriers; pedestrians are natural Pythagoreans, preferring the hypotenuse to the other two sides of the triangle wherever possible. (Adams, 2004: 39-40)

Adams' Pythagorean pedestrians are not just concerned with saving time; they are also concerned with saving energy. Pedestrians and cyclists are unique in the urban environment in that they move under their own power and will often take the shortest and (particularly for cyclists) flattest route (Gehl, 2001; Parkin et al.

Figure 9.2   'Pedestrian rhythms' (author's collection)

2007). However, as Adams (1995, 2004) asserts, since the urban environment has been planned as an isotrophic surface, human energy expenditure has ceased to become a variable because it is erased by the use of mechanical and electrical locomotion. Indeed, as Bendixson (1974) points out, there has been a failure to take into account any kinaesthetic dimensions of movement:

> The first consideration has been to prevent pedestrians from getting in the way of vehicles (the safety of bicyclists has been ignored). The psychological and physical requirements of people on foot have been disregarded. (1974: 57)

During my fieldwork I documented countless examples of the energy-space of cyclists re-interpreting the rhythmic score of the city. Near the end of Zara's journey to work for example, she encounters a one way system en route to the Pentonville Road from Farringdon, where she elects to ride on the pavement:

> The first few times I did that (section of pavement) I did get off and walk but then I just got bored, and thought well if I go slowly, what's the difference? [...] The reasoning behind going up on the pavement there is that otherwise you have to go all the way round the whole Kings' Cross one-way system which adds probably at least five minutes on to the journey. Kings' Cross the traffic is

> a nightmare, and you know everyone is all over the place and you have to go
> right the way round and go through tons of sets of traffic lights, then cross over
> Pentonville road which is a busy road. It was like the first time I did it I just
> thought this is stupid, I'm wasting my time and energy and so the second time I
> went out like that. (interview, 1 November 2004)

The layout of the one-way system that Zara describes promotes a complex and continuous rhythm which keeps traffic moving but over a spatially extended and indirect route. As a cyclist however, Zara interprets this extra distance as additional energy expenditure. Consequently, she chooses to ignore the long-winded staccato rhythm embodied in the one-way system and with the embodied experience of both time and energy to the fore, instead improvises a slower but more direct rhythm which has more in common with pedestrian movement. What she implies is a profoundly contextual understanding of rhythm rooted in the conjoining of the bike and body. Zara's route suggests an embodied and situated understanding that a slow, continuous and direct rhythm will use less energy than a long stop-start rhythm and therefore be more appropriate for the instrument she is using and what she wants to achieve.

Affective rhythms: vulnerability

> Lights aren't meant to be for pushies anyway, they're just to keep the traffic
> under control. (Elina, 11 July 2005).

Risk as both Adams (1985, 1995) and Rowe (2004) remark, has different meanings for different people. However, as Adams observes, 'traditional highway engineering has been based on the theory that we are completely oblivious to dangers in the environment around us' (cited in Clarke, 2006: 291). One of the main problems is as Fischer (2003: 423/4) points out, because the quantification of risk is largely technical in nature and sees social factors as irrelevant, its assessment is 'sociologically uninformed' and there is a 'reluctance to consider contextual factors directly'. Consequently features such as traffic signals are designed to manage the level of risk at particular localities. In doing so they impose a particular order and rhythm on road users, dictating when users should stop and go whatever the specific circumstances. However, as Rowe (2004) points out,

> risk is not a thing nor is it an attribute of design or of public spaces. The
> mathematical probability of an event occurring can be calculated, but individuals
> then decide what that probability means to them. 'Risk' is an abstract noun
> created to cover the multitude of ways in which individuals can perceive a
> situation in relation to their own personal safety. (2004: 14)

Congruently, all of the cyclists I worked with – whether they obeyed laws or not – were profoundly aware of their own vulnerability. Without the protection that a

Figure 9.3   'Noam pushing the beat' (author's collection)

ton of metal affords the car driver, riders displayed an embodied awareness of their own fragility and consequently rode in ways which could minimise the danger to themselves based on their perception of the immediate context.

As a result, cyclists can frequently be observed disobeying the material and socio-legal directives to stop at junctions because of situated understandings of their own personal safety and subsequent attempts to minimise conflict with vehicles that could potentially injure them. Certainly, one of the most widely cited reasons by cyclists for the seemingly irrational behaviour of disobeying traffic signals is somewhat counter-intuitively that they deem it safer to do so. This point is emphasised by Noam in his account of jumping a red light:

> Yeah I do that (jump the lights) because if I stay on the other side I get stuck, so it makes sense to do that. Otherwise you have to stand in the middle of the road with cars going on either side of you waiting for them to stop so you might as well get across the moment you can so you don't have to do that. It makes a lot of sense in terms of my personal safety. I mean there are a lot of things that you do on a bicycle that are not in The Highway Code but make a lot of sense for one's safety and you know The Highway Code is designed to ensure everyone's safety but it doesn't for cyclists. It's ensuring that car drivers don't run over pedestrians; as a cyclist you're just out there. (Noam interview, 22 November 2005).

The 'appropriate' rhythmic performance at such junctions would be to stop and wait at the light, moving off when it is 'safe' to do so. However, in this instance Noam slows, looks and continues whilst the light is still red, imposing a new rhythm at the junction; pushing the beat of the official rhythm, a decision based largely upon his embodied experience as a vulnerable road user and the fact that he feels uncomfortable moving off with the other vehicles. Noam argues that this difference in vulnerability is not accounted for in the design of the junction timing or in the socio-legal framing of The Highway Code. His account of negotiating this particular junction suggests that risks are purely contextual and contingent on situated and embodied knowledges. Thus where system design fails to account for the affordances of a particular user, as Fischer (2003) acknowledges, seemingly 'risky' behaviour can often be a fundamental source of safety. Again however, the ability to perform such 'risky' rhythms is seen to be framed by the disposition of the rider and the skills that they possess. Without these, a sense of vulnerability itself is not necessarily enough to afford the possibility of new rhythms, for other factors must also coincide with these situated and embodied understandings in order to open up new 'lines of flight'.

Conclusions

This chapter has demonstrated the inter-relatedness of the material and immaterial to any understanding of how urban places are experienced and contested. I have shown how the materialities of the bike and the built environment intersect with and produce the immaterialities of vulnerability, time and energy which are in turn productive of rhythmic variation in the city.

Rather than simply reproduce the dominant rhythms appropriate to these spaces, the accounts here suggest that the personal, situated understandings of riders are central to the construction of different readings of time and space. However, such unprogrammed uses of space fall short of the 'ideal movements' designated as legitimate by the system's designers and are often deemed illegal and criminalised, constructed as irresponsible and dangerous by many commentators (Fincham, 2006; Horton, 2006). Such readings have not been helped by accounts which theorise these unprogrammed uses of space as resistant to official rhythms. I argue however that the rhythmic variations outlined here are neither always 'criminal' (though they may be criminalised) nor resistant; rather for many riders these actions constitute the rational responses of people performing a rhythm written with other 'instruments' in mind. Whilst these accounts identify different rhythms, they do not suggest the rejection of the dominant rhythm; rather there is an over-laying of a new rhythm which syncopates with other rhythms.

As I have already noted, I do not want to impute a homogeneity to cyclists for not all are the same. Whilst I have largely theorised rhythmic variation in relation to the differing technologies and affordances of the car and the bike, many other factors direct an agents' ability and willingness to interpret the score differently. As with any musician, the possible rhythms that can be played are not only influenced by the choice of instrument and the context of the performance, they are also influenced by the skill and disposition of the performer. I suggest that this may be a key reason why official and appropriate rhythms are often reproduced: many riders lack the bodily ability or desire to perform new rhythms, or perhaps have been more strongly trained to adhere to the socio-legal discourse of the bike-as-car. As a result and despite the use of an 'inappropriate' instrument these riders may well reproduce the vehicular rhythms embodied in the built environment. Whatever the case, as Edensor notes, the heterogeneous movements of cyclists represent just one interaction amongst many intersecting practices in any social space where multiple rhythms collide, syncopate and synchronise to produce dynamic spatialities (2009: 2).

Chapter 10

Repetition and Difference: Rhythms and Mobile Place-making in Santiago de Chile

Paola Jiron

Introduction

As the location of particular sets of intersecting social relations and activity spaces (Massey, 1995) places can be characterised as open, permeable and always in construction, unbound and mobile. Places can be constituted through reiterative social practices, including the practice of urban daily mobility, presented here as crucial to analysing contemporary urban living. As socially produced motion, mobility implies giving social meaning to the practice of moving from one place to another, and it involves an appropriation and transformation of the space encountered during this practice, thus generating mobile places.

The reiterative characteristic of places, along with the moving characteristics of mobility practices, suggests an investigative approach to mobile place making by exploring the multiple and diverse rhythms present in this practice. Rhythms are pluralistic, diverse, relative and repetitive; this and particularly their lineal/ cyclical repetitiveness, makes them inextricably linked to time, just as important as the appropriation of space in place making.

Using an ethnographic approach to urban daily mobility practices in Santiago de Chile, this chapter analyses the everyday generation of mobile places. It argues firstly that place making can be generated on the spaces encountered in mobility, that is those spaces travelled on, in, by, through or within: buses, Metros, cars, bicycles or foot, become mobile places. Secondly, it discusses how in the analysis of mobile place making, linear and cyclical rhythms can offer opportunities for individual rhythms 'of the self' to avoid being subsumed by the rhythms 'of the other' (Lefebvre, 2004), thus emphasising the multiple possibilities for daily travellers to 'tailor their journeys in accordance with their own strategies, imperatives and feelings' (Edensor, 2009).

Mobility and everyday practices

A major aspect of what makes mobility significant is that it greatly impacts upon people's daily life; lives do not stop while being mobile. The time spent travelling is not 'wasted' (Jain, 2006), and much occurs during these mobile moments. Although mobility has mainly been conceived as physical, it can also be virtual or imaginative (Sheller and Urry, 2006; Szerszynski and Urry, 2006), and the use of television, the internet or mobile phones allow for the possibility of being present in more than one place at the same time. These technological advances make the management of distance increasingly relevant and the choices of presence can be broadened to three: co-presence, which eliminates distance; mobility, which handles it through displacement; and telecommunications, which transfers dematerialised information (Bourdin, 2003). However, despite constant innovations in technologies, being physically present remains imperative in daily life and physical travel is still required to make necessary connections. According to Urry (2004), virtual and physical travel transforms the nature of and need for co-presence; thus virtual travel complements physical travel which in turn leads to co-presence. Urry (2003) also asserts that it is unlikely that virtual communications will alter the importance of meeting or face-to-face connection, hence the need to look at how these moments of connection and encounter occur.

Kaufmann (2002) explains issues of co-presence in terms of connexity and contiguity. Connexity can be defined as establishing relations using the intermediary of technical systems, whereas in contiguity this relationship is established by spatial proximity, implying density. Connexity cancels out spatial distance to allow for actors' interaction through trains, aeroplanes or automobiles. Kaufmann explains that these are often thought of as having a tunnel effect, meaning that whatever lies outside these modes is invisible, avoided, ignored, and that 'the appropriation of space crossed between the origin and the destination is not possible' (2002: 23) due to speed.

However, as will be discussed in the next section, I argue that the space inside can indeed be appropriated and signified; that is, the train, aeroplane or automobile can become meaningful spaces in themselves. It may appear that, as connexity increases territories lose importance, but in practice contiguity and connexity may have a tendency to merge. These spaces generate experiences which may be characterised as being reversible or irreversible. Kaufman (2002) understands reversibility as the impact mobility has on actors' identities, that is, the traces left behind in the body and consciousness by mobility experiences. A reversible experience would not leave any trace behind, as it would quickly disappear, if ever present; whereas an irreversible one would remain in the person's body, emotions and consciousness. Reversibility is linked to the process of place-making as temporal and spatial experiences may become impregnated in a person's body or memory, thus modifying the relation with space.

Places, spaces and time

Under current global processes, some believe that spaces lose their distinctiveness and become subdued and unified, making place lose its significance with its characteristics emptied and abstracted (Harvey, 1996); others insist that place persists as a constituent element of social life and historical change (Gieryn, 2000; Sheller and Urry, 2006). According to Savage et al. (2005), while still relevant, the process of place making in contemporary cities is complex, and people's sense of place making changes and is not so much based on historical attachment but on choice. Massey (1994, 1995) has argued that if the social organisation of space is changing, disrupting existing ideas about place, then the concept of place should be rethought as the location of particular sets of intersecting social relations and activity spaces in time.

This chapter develops this re-conceptualisation, thus place refers to a location, a locale and meaning (Cresswell, 2004; Agnew, 2005) which involves an appropriation and transformation of space and nature, inseparable from the reproduction and transformation of society in time and space. In this progressive sense, place is open, permeable and always in construction, and is constituted through reiterative social practices, such as the everyday mobility practices discussed here, which remake place on a daily basis. As Cresswell (2004) mentions, places are never complete, finished or bounded, they are always becoming, in process.

Due to the multiplicity of changes in space and time in terms of speed, forms and encounters, Massey (2005; 2007) refers to places as events, a constellation of trajectories and processes, multiple and not necessarily coherent. In these events, multiple temporalities collide, synchronise and interweave (Crang, 2001). The event of place requires negotiation and poses a challenge as to how encounters with others (or things) will take place. As events, places cannot be predetermined or anticipated; they occur as they happen in time and space and although certain regulations may produce situations thato repeat steadily, a similar constellation is difficult to obtain on a daily basis.

Hence, places are about relationships, about placing (or displacing or replacing) people, materials, images, and the systems of difference they perform (Sheller and Urry, 2006). This means that places are not experienced in a similar manner by everyone, for place is both the context for practice as well as a product of practice. The relationship between places and practices, particularly those which occur on a daily basis, are productive of contemporary urban life.

Practices of mobility, whether through travel and tourism, migration, residential mobility or everyday practices characterise modern living; the latter being the focus of this research. Place making occurs in fixed spaces, and also on the spaces travelled on, by, within. Understood as socially produced motion, mobility implies giving meaning to the practice of moving from one place to another and suggests the possibility of places being appropriated and transformed during this practice, generating what I term 'mobile places' and 'transient places'. Mobile places are those that people signify while travelling on and in them: cars, buses, metros, trains, or bicycles. In transport and urban planning the time spent on these is often represented as 'dead time' (Urry, 2006; Jain, 2006) and policy interventions are aimed at diminishing travelling time and improving connections by making it more efficient. As will be seen in the narratives in this chapter, travel time is experienced differently by different people, and not everyone experiences it as dead time; on the contrary, for many the moments spent on different transport modes are crucial to their everyday existence.

Transient places involve those fixed spaces which people signify while moving through them. They are not places of permanence but places of transit and transition elsewhere. Regardless of the amount of time spent travelling through them, they are nonetheless appropriated and signified. These often vary in type, form and permanence and are sometimes understood as public spaces or spaces of public use. Those most commonly studied include markets (Cresswell, 2006), bus stops, petrol stations (Normark, 2006; Sabbagh, 2006), airports, parks, and streets (Duneier, 1999). These have been conceived as 'non-places' (Augé, 1995), spaces of institutions 'formed in relation to certain ends (transport, transit, commerce, leisure)' (Kaufmann 2002: 94). Augé does not use this term in a derogatory way, but rather to describe a certain sort of place that inculcates a new sense of thin or abstract identity. Relph, on the other hand, points to such places as strip malls, new towns, international architecture style and tourist landscapes as examples of contemporary placelessness (in Agnew 2005). However, as Agnew (ibid.) argues, 'placelessness' is in the eye of the beholder, as malls, markets or bus stops are not just spots along the way, but reflect important meanings to people's everyday experiences. This chapter focuses on the first type of places: mobile places.

Rhythms and mobile places

The dichotomy, overlap, similarity and juxtaposition of space and time have been a major topic of discussion in geography and urban studies. Torsten Hägerstrand's (1970) time-space geography pioneered the study of sociospatial analysis, asserting the indissoluble link between time and space. The notion of 'timespace' developed by May and Thrift (2001) is also helpful in apprehending the mobility turn insofar as it relates to the interconnectedness of time and space, and the multiplicity of timespaces. It attempts to overcome the dualism in understanding time and space as separate entities and sees them instead as analytically inseparable. Timespace and its experience, as a multidimensional, uneven and always partial process, becomes pertinent in the context of mobile urban life, since changes in the nature and experience of either space or time impact upon changes in the nature and experience of the other.

Additionally, the experience of social time is multiple and heterogeneous and varies both within cultures and between societies and individuals, being related to their social position. May and Thrift identify four interrelated domains where time and space have particular implications for social practices. In the first, the experience of timespace varies according to timetables and rhythms, according to daily cycles, seasons or body rhythms. In terms of mobility, this may have differentiated impacts on social practices: people may use different modes of transport according to seasons or access to the places to and from which they travel may be highly dependent on the times of day. For the second, timespace is also shaped by systems of social discipline – or Lefebvre's linear rhythms (2004) – including work time, home time, religious time and leisure time which have different meanings in different spaces. These timespaces influence how, why and where people move at certain times; for instance, the experience of time for going to work is different from the time for going to church. A third domain concerns the relationship with instruments and devices which affect the way time and space relate with social practice. Mobile phones or computers have made physical mobility less necessary at times, and cars and rapid forms of transportation have major implications for the way timespace is experienced. Lastly, May and Thrift point to the ways in which timespace is translated into various forms of representation. Where this involves the study of mobility, it requires understanding travel patterns, travel experiences, travelling conditions or the consequences of mobility, along with other practices that take place in timespace. Consequently, it also involves creating new methods to capture these mobile practices in timespace and representing them in innovative ways, including methods like the mobile ethnography presented here. By using such approaches, the picture that emerges is not that of a singular or uniform social time stretching over homogeneous space, but rather that of various (and uneven) networks of time stretching in different and divergent directions across uneven social space.

The multiplicity of timespaces can be best understood using Lefebvre's (2004) work on rhythmanalysis, which attempts to deal with the temporal orders of everyday life (Meyer 2008) and unifies time and space through the analysis of rhythms. Lefebvre understands rhythms in terms of a relational repetition, meaning that there is always a certain comparison to other rhythms: slow, varied, repetitive, long, empty. Although rhythms are repetitive, they are not necessarily similar; in their repetitiveness, difference is always found as no rhythm is ever the same as another, and every rhythm has its own beat. In this repetitiveness, one of the main distinctions made by Lefebvre involves the difference between cyclical and linear rhythms.

The first refers to the natural and cosmic cycles of the universe that human beings are exposed to, including the recurrence of days, weeks, years, seasons and tides which have no beginning or end. They are repetitive cycles but always different in their repetition. Linear time, on the other hand, emerges from social practices, particularly those of work, and refers to the 'monotony of actions and of movements, imposed structures' (2004: 8). Lefebvre suggests an antagonistic unity between cyclical and linear rhythms which are nevertheless inextricably related, in perpetual interaction, exerting a reciprocal action on each other: 'everything is cyclical repetition through linear repetitions' (8). One becomes the measure of the other, making their interferences particularly relevant to explore.

In the analysis of rhythms the body becomes essential, including the polyrhythmic and eurhythmic body. According to Lefebvre, in rhythmanalysis, the body is crucial because of its unique rhythms and its capacity to perceive the rhythms outside of it. This is particularly relevant in the process of mobile place making, as through all its senses, the body perceives movement and experience places in of all sorts of ways. The senses used in the process of place making as well as the way social characteristics are reflected on the body, such as the colour of skin, disability, being young, old, pregnant or blind, affect mobility practices.

Following the work of Crang (2001), who highlights the importance of the multiple rhythms and temporalities of urban life and uses time geography to analyse daily trajectories, and Edensor and Holloway (2008), who show how rhythmanalysis can highlight the experience of mobility in space and time, this research analyses how rhythms can help to understand mobile place making.

The following sections present the results from a mobile ethnographic study carried out in Santiago de Chile. The narratives examine processes of mobile place making through the rhythmic mobile practices of two urban travellers, Roberto, who takes the metro everyday and manages to find moments of place-making in the ride; and Isabel, who, at 80 years of age, thoroughly embodies the rhythmic travel experience on the metro. Though Lefebvre suggests going outside of rhythms in order to grasp and analyse them, both travellers were shadowed during their mobility practices.

A place for reflection: the Metro

Santiago's Metro system started running in 1975, and its lines have been continuously extended since, with five under and overground lines running North, South, East and West of the city. It is considered an important symbol of Chilean modernity due to its speed, efficiency and cleanliness (Procalidad, 2002). In the midst of contemporary harried lifestyles, the Metro's reliability and comfort make it a refuge for some, or a place for reflection and introspection. The following experiences of Roberto and Isabel illustrate this.

Roberto is 42 years old and lives with Cecilia, his second wife, in Jardines de la Viña, a middle income neighbourhood. At the age of 25 he married for the first time, and soon after separated. Unlike Cecilia, Roberto never went to University. He is an accountant by trade and his lack of education greatly hinders his chances of a more stable job and better salary. He works at an exporting company on the North-eastern side of the city, close to the airport and is not overly excited about his job. He works very long hours, including Saturdays. The taxi/metro/bus ride across the city lasts about 2 hours each way. His days are controlled by the linear rhythms of the city and his job. However, glimpses of his own imperatives and feelings emerge during his journey.

His journey starts at 6:40 when he walks out of his house towards the taxi stand. At this time, the queue is short and soon he's on his way to the Metro. Once embarked, he relaxes, stands by the window and contemplates the scenery (Figure 10.1). It's his moment, the only timespace in the day when he's alone and gets to think, to go over his thoughts, his days. He manages to break from the structures of the journey. He changes lines and complains about the excessive advertising; he says he feels invaded by it (Figure 10.2).

Figure 10.1   Morning view from inside Metro

He likes the Metro, he says, has been riding it since it first started and knows something particular about every station: the one with the mural, the one with the paintings of the ocean, the one with the tiles painted by children... (Figure 10.3). He remembers when each station used to have an icon identifying it.

During the journey, he never sits down, doesn't talk to anyone, and doesn't make eye contact. For him, it's like being in his own world. As the wagons get emptier he begins to feel more comfortable. He used to read books in the Metro, but not anymore. He says, 'I'm too tired, I now pick up Publimetro or La Hora [free newspapers] and I read it all, I read every detail of the news or do the crossword puzzles or sudoku, I really like them'.

He gets off the Metro and changes to a private bus that takes him to his job at ENEA, an industrial park, very close to the airport. Most of the area is still under construction, making the surroundings of his office building quite bare. He feels isolated at work, as it's difficult to leave since buses reduce their frequency after 9:00 am and travelling rhythms soften as workers remain indoors. He has lunch taken to his office and does not leave the building until 20:00 or 20:30, or sometimes later, when he returns to catch the bus that takes him to the Metro.

Once on it, Roberto is not oblivious to the outside scenery. For him, riding overground is the best part of the journey. 'On the way back, I don't read anything,

Figure 10.2   Advertisement inside Metro

I just contemplate'. He says it changes every day. 'I like the Metro coming back, it goes overground after Ñuble station, and I enjoy it because you can watch the scenery, although it's the same route, it may not vary one centimetre, to the right or the left, but it does change. It doesn't matter if it's dark, you see the lights. I like winter better than summer, I like it because I watch the clouds, the rain, the storms, the wind, the snow on the Cordillera is absolutely beautiful, or the sunsets (Figure 10.4). I use that time to think, to look, to distract myself just by looking at the scenery. Think about it, you are in the Metro and can somehow appreciate the outside while inside it, it's amazing. For me it's not really lost time, I use that time to think, it's time for me. It's nothing special, but I kind of renovate myself, I start rewinding the film. I like the time on my way back on the Metro, although it's the same, the outside changes […] When you're inside the tunnel it's a drag, plus at the times I travel it's very compressed and uncomfortable; that's the bad part of the Metro, the quality of people, you start losing the quality of people,

Figure 10.3   Painting inside Metro station

Figure 10.4   View from Metro: Cordillera de los Andes

independently of the person next to you, they don't care if they have a big bag that squeezes you, they don't care, it's practically like a tin of sardines at times, but I still like travelling on it'.

At first glance it appears as though Roberto endures the daily journey, for a two-hour journey to cross the city seems tedious and extremely long. However, when analysed in the context of his broader life, the linear rhythms he follows, the demands and frustrations of work and family, it appears as though his time on the Metro is the only timespace he has to himself, his timespace for contemplation. He explains about the demands made by his wife of needing more time, more study, more money, more babies, and a better lifestyle; at work he complains about the mediocrity of the people he works with, the long hours and low pay, because he does not have the necessary academic credentials to be properly recognised for the job he already does. He gets home late, works all day Saturday, and Sundays are the only days he can rest, be with his wife and undertake the family events and chores required.

His travel time is the only 'in between' time to think, dream and drift away. His attitude on the Metro is a *blasé* one (Simmel, 1969), purposely ignoring the rest of the passengers and feeling annoyed by those who interrupt his thinking. His body is stiff, standing by the window and just contemplating. The Metro is not just a mode of transport from home to work; it is a mode of transport to drift away, to the times and places where he will never be. This 'in between' timespace is his way of resisting the rhythms of others and imposing his own.

Travelling by Metro: A place for independence

Isabel is 80 years old; she lives in Jardín Alto, a lower middle income neighbourhood, with her daughter and her daughter's family. For many years she lived in Sweden as a political refugee with her family, most of whom have returned to Chile since the return to democracy. After leading a life of political commitment and social awareness, and exploring new worlds, she returned to Chile to attempt to continue living as independently as when she was living abroad. When she first returned she lived alone, but her daughter soon insisted she lived with her, mainly because she was getting old.

Every Wednesday she visits her best friend from childhood, Soledad, who is 81 and lives in downtown Santiago. Although Soledad's son takes her to the doctor and she has someone helping her everyday, Isabel visits her once a week to run errands with her to the pharmacy, the supermarket, or just for a coffee close to home. For both of them this is their weekly outing and a routine they have been doing for years. Isabel appears to be in very good shape, but she says she feels increasingly tired, she can't see very well and has trouble going up and down the stairs. She has made herself a weekly routine which she follows strictly. However, when things change, like a new bus route, or a new Metro station opens, it's dreadful for her, as changes in city rhythms means finding her way around anew and she has difficulty. This once-a-week outing is her way of still feeling useful, alive and independent.

Figure 10.5   Isabel climbs the Metro stairs

She tries to avoid hectic rhythms and waits until the city calms down. Around 11:00 am, she strolls down the street to catch the taxi that will take her to the Metro. Although she sits in the front seat, the car is old and uncomfortable and she has trouble getting up from the seat and seems disoriented crossing the street to enter the Metro station. Once she reaches the Metro, she feels at home, she knows it well, knows her way, where to go, how it works. She has become used to it by now; she has a close understanding of its rhythms, hence she avoids busy times. Security guards are always helpful, she says. Although she has trouble going up the stairs, she grabs the hand rail and goes up slowly (Figure 10.5). She knows which way to turn and where to wait for the appropriate wagon. Once inside the train, she takes a seat, changes her glasses and enjoys the ride (Figure 10.6).

She loves the Metro; each station has a special design and generates a special meaning to her. Her favourite one is the one with the palm trees. She says, 'you know, the one with the long palm trees on the walls, it's so nicely designed, they really thought about it, the whole station is green and the palm trees just emerge from the ground, look' (Figure 10.7). In Baquedano station she changes to Line 2 and it is slightly difficult for her as there are many stairs and it is more crowded. People are not always kind on the Metro, but still one person gets up to offer her a seat. She then reaches her station and gets out. She gets scared of going back on the Metro on her own at night, particularly because the taxis are not very regular then, she says, and she fears getting off at the wrong station and having no way to get back home. So today, her daughter is picking her up.

Figure 10.6   Isabel inside the train

New stations are difficult, 'specially now, for me Vicente Valdés [station] is a mess, because I am still trying to adapt myself to those new levels, because they have two or three levels, the lifts and mechanic escalators are located only in some places within the station, so sometimes I'm in the lift and I don't know which button to press. There is something written, but I don't see too well, I'm almost blind, with these [glasses] that have so much prescription. I see awfully, and the other dark ones are worse, so if I want to see the sign I confuse the letters, and don't see well, so I end up asking people which button, where do I have to go, and it's embarrassing'.

As an embodied experience, mobility is specifically sensitive to the way the body performs in mobile places. The body is sensitive to the rhythms lying outside of it, the multiple and diverse rhythms that are captured by the senses, and also performs in accordance with the various rhythms and situations it faces. Isabel uses the rhythms of Metro and public transport to suit her needs to feel independent and

Figure 10.7   Tiles in the Metro station

still capable. She refuses to be a burden on her family yet understands that she no longer has the agile body she used to have in order to explore the streets. Her sight is weak, her strength frail, yet her she finds the energy to continue repeating the weekly practice of going out to visit her friend.

This weekly outing downtown does not only involve the experience of socialising with a friend, but is provide the experience of being capable of undertaking the journey that takes her there. She skilfully appropriates the space of the Metro with her aging body. Though the Metro's infrastructure is not always prepared or equipped to accept this body, she nonetheless makes it suit herself by making it her mobile place, where she strolls independently, at the most convenient times and days that are easier for her, on the routes that are most familiar, that she has made and remade over time. She admires the spaces she goes through and the whole experience has meaning and remains with her for at least a week, making her look forward to the next occasion.

At the age of 80, Isabel's weekly outing is one of her only opportunities to connect to the outside world, to feel useful and independent. However, the infrastructure available is not designed for her needs: taxis are uncomfortable as she has difficulty bending down to get in and out; bus stops are unsafe and uncomfortable as they are unprotected from the weather and there is no place to sit, bus drivers are careless and rough and schedules are unreliable; the Metro's constantly changing operation confuses her, signposting is too small and not easily located, their are many stairs to climb, and it is easy for her to trip with loose surfaces or holes on the badly maintained sidewalks. Finally, lighting is inadequate at night, making her scared of being in public areas. One of the most difficult problems is that of unfriendly passengers, who do not respect the elderly and their travelling difficulties. Her best consolation is the Metro security guards whom she seeks when she needs help.

Both travel experiences featured above accompany passage through different stages of life, different stages in a cyclical rhythm, which affect the appreciation of time and space, and thus the process of place making. Roberto is in his 40s, married, already has a child and is expecting another, and his education and preparation have already occurred. He is somewhat resigned to his current situation and it appears as though his daily routine leave him drained. However, his approach to daily travel is introspective, following his own rhythm: it is his time to think and be with himself, away from all the responsibilities that surround him. That is why he ignores others, as he mainly cares about the spaces he encounters.

Although Isabel is in her 80s and has lived a very hectic, fruitful and difficult life, remaking her life elsewhere, she has managed to reconstruct it in Chile. She compares the spaces she encounters with other places she has been to and sees the problems in the city as a reflection of society. However, she still feels useful and dares ₊o go out and see what is out there, even if it scares her; she manages to contro¹ ₊ne spaces she goes to so she can still enjoy her weekly outing.

S᾽₊adowing these experiences is important because from an outsider's pᵉ ₊pective, the mobility experiences might appear as reversible, meaning that ₊ew traces of them remain after the journey due to their apparently monotonous and meaningless use of time and space. However for these participants, although their journeys are part of their daily (or weekly) routine, they do not lose track of the characteristics of place; they are familiar with them and become part of them. The spaces encountered on or by the Metro are signified along the way and each traveller uses the rhythms of the city as best suits them: Roberto finds his way of drifting away from the linear rhythms while Isabel takes advantage of these rhythms for her own personal strategies.

Conclusions

The narratives presented above provide insights into understanding how place making is possible in everyday mobile practices. Through mobility, place-making is generated while travelling on moving objects such as a train, bus or car. While urban travellers may not experience these spaces in a similar manner, they are significant to some as a place of contemplation, reflection, independence and socialisation. The possibility of signifying these timespaces exists by dodging or understanding rhythms of the other, and imposing rhythms of the self. Practices of moving on, through or by those spaces may thus enrich people's urban experiences, making them valuable and irreplaceable.

According to Agnew, for humanists, 'places are woven together through space by movement and by the network ties that produce places as changing constellation of human commitments, capacities and strategies' (2005: 90). Although this is true, mobility does not just bind or weave places together: places are created *in* mobility and *through* mobility, not just in spite of mobility. As seen here, the act of moving or repeating mobile routines creates mobile places that are meaningful. For the travellers presented, the Metro is a concrete significant space, a place they appropriate, look forward to getting to in order to unwind and prepare for another phase of their lives. It is an intermediate, 'in between' place where they stop having to perform as they move from a job to home where they have other responsibilities. Roberto has a introspective experience; he drifts to another world in the Metro, puts order into his life, and enjoys the beauty outside. For Isabel, by contrast, the experience of riding the Metro leaves with her a sense of being capable and independent in the space encountered.

In the context of mobility practices, co-presence remains significant, but so are the moments that people are capable of being isolated within conditions of co-presence, are able to be alone in the midst of multitudes. It seems that in the context of contemporary harried lifestyles, with a multiplicity of demands, and increasingly structured rhythms which regulate daily life, travel time offers an opportunity for co-presence, but also of stopping to take some air, to revise thoughts and be alone. Mobility appears to be producing in-between spaces that might otherwise have occurred at home, and people are increasingly providing meaning to these moments and making them valuable. Through choice or not, long periods on the move have caused people to adapt to rhythms, sometimes significantly improving their lifestyles, or conversely, impose their own rhythms in breaking away from increasingly demanding lives.

Chapter 11

'The Engine Sang an Even Song': Rhythm and Mobilities among Early Women Aviators

Dydia DeLyser

In 1929, record-setting pilot Ruth Nichols was completing a 12,000-mile US air tour when she was invited to race in the first Women's Transcontinental Air Derby. At that time there were only about 100 licensed women pilots in the US, and of those, few had the kind of flying time required to qualify for the Race – just qualifying was a significant accomplishment (see Nichols 1957; and Brooks-Pazmany 1983). Nichols had been flying the air tour in a Curtis Fledgling, what she called a 'tried-and-true old friend,' but it was a light airplane that would not be competitive in the Derby – 'its plodding pace,' she wrote, 'could not compete with the far higher-speed aircraft that would be entered in the race' (1957: 78). Eager to compete, she hurriedly arranged to borrow a high-powered Rearwin aircraft from the President of the Rearwin Company. She flew the Fledgling to Saint Louis, and there picked up the Rearwin to fly it across the country to Santa Monica, and the start of the Derby. As she flew, the Rearwin's 'motor sang sweetly' (1957: 79). Though she was unfamiliar with the aircraft, having never flown it before, she enjoyed it immediately: 'The sleek red Rearwin plane was a joy to handle after so many months at the stick of the lumbering Fledgling,' and she described her experiences and interactions with the airplane in detail: 'I reveled in her speed and slid light-heartedly across the Kansas sky. I felt like a queen bee, speeding gaily and regally through space ... .' (ibid.). But such interactions were not always joyful: 'when I approached Whichita, making knots on a long power descent, suddenly the motor quit cold' (ibid.). Now, she wrote:

> With only three hundred feet of altitude there was neither time nor distance to be choosy about a landing spot. My mind sped as follows: Only small fenced-in areas on all sides ... Must stall-in over fence of tiny pasture dead ahead. . . . Better slip and fishtail ... Creepers! What a floater! [Meaning that the airplane was flying so well it didn't yet want to land.] Running out of field and the wheels haven't touched yet. ... Full brakes, she'll turn over ... Better just hit the fence head-on and pray ... Duck! ...
>
> R-i-i-ip! ... Jolt! ... Silence.
>
> I lift my head. I'm still alive! The plane is right side up—in the next field'. (80)

Climbing out of the airplane she saw she had managed to sail across two ditches and had broken two fences, 'carrying wire and fence poles along' with her. Inspecting the damage she found 'Wings, landing gear, motor—all in tact' and that there were only a few 'tears in the fabric on the underside of one wing.' Then, upon further inspection she found, much to her embarrassment 'that the motor had quit merely because it was out of fuel, and that all the time I had had a full reserve tank of gas. That was what came of hurrying too much' (1957: 80).

This chapter, about the rhythm of flying, and what happens when that rhythm is broken, explores the mobile and embodied interactions of 1920s women pilots with particular rhythms, those of their flying and of their aircraft engines. These rhythms, often joyfully acknowledged though never unnoticed when they were right, could become a harbinger of doom when they were not. In the air, and with little time to spare, women pilots responded to engine arrhythmias as Nichols did: in a flash of both conditioned and spontaneous embodied responses, responses so quick they were unreflective in the moment but so important they were thought about long before and long afterward. This chapter, part of a larger research project on issues of gender and mobilities among early women pilots, examines the ways that women's aerial mobilities were rhythmically both enabled and constrained, and suggests connections for the ways geographers understand the rhythmic and embodied experiences of mobilities.

Rhythms of aerial mobilities

In order to gain insights into the affective and emb ɹied rhythms of flight I engage with the recent upwelling of interest in mˠˈ.nties, called by some, like Kevin Hannam, Mimi Sheller and John Urry ˈ.ɹ 'mobilities turn' (2006) – a broad array of research that includes interestˢ ᵼ both the material aspects of movement and transportation, and the ways ᵗ ɹse movements are socioculturally and sociospatially understood (Cresswel¹ ˉ.ɹ06). Significantly, where in traditional transportation research, scholarˢ ,ɹsited that travel was about reaching a destination, and that time spent tra ɹɪng was wasted time that travelers sought to minimize, mobilities scholars seek also to understand (among other things) the embodied nature and experience of travel and mobile experiences themselves (Cresswell 2006; Sheller and Urry 2006; Bissell 2007). For as Tim Cresswell has explained, mobility must be viewed as more than 'getting from A to B,' and seen to include the ways that different embodied mobilities can be practiced, contested, represented, and understood (2006: 265). Further, such mobilities, as Sheller and Urry point out, must be understood as 'hybrid systems, 'materialities and mobilities', that combine objects, technologies, and socialities' along with the people who engage them (2006: 214).

For early women pilots like Ruth Nichols, the journey itself was the point, and their interactions with their aircraft shaped what those journeys and those mobilities would become. I attempt to forward discussions about mobilities specifically by thinking through such embodied interactions with objects and technologies. I explore how women's aerial mobilities were shaped by rhythms and arrhythmias, because, as Tim Edensor and Julian Holloway put it, 'Rhythm is intimately associated with movement' (2008: 488), and therefore understanding rhythms is part of understanding mobilities. Still, as Justin Spinney has observed, though rhythms are 'intrinsic to movement' they are 'not innate: certain conditions must be met to maintain rhythms; technologies and bodies must be honed and maintained to enable continuity' (2006: 718). To examine these ideas, I engage the rich autobiographical accounts of two record-setting pilots of the late 1920s and early '30s, Ruth Nichols (1957) and Louise Thaden ([1938] 2004).

Interestingly, within the mobilities literature, and even within the more specialized 'aeromobilities' literature, scholars have overlooked *flying* (by which I mean actually piloting an aircraft), and what attention has been paid to travel by air has tended to view it from the perspective of destinations, or of airline passengers as 'one space of flows that increasingly moves people apparently (though never actually) seamlessly around the world' – an approach that obscures the embodied experiences of flying, and the pilot's experiences of the journey (Sheller and Urry 2006: 219; see Vowels 2006; Adey 2008). True, airports, air transportation, air travel, airlines, air traffic and air power have all been examined from an array of different perspectives (see Gottdeiner 2001; Pascoe 2001; Cwerner 2006; Kaplan 2006; Lassen 2006; Adey 2008). And neither is it as if the embodied experiences of mobilities in general have been ignored: scholars have explored dancing, walking, driving, hiking, cycling, and waiting in insightful, experiential detail (see, for example, Morin 1999; Featherstone et al. 2005; Merriman 2005; Wylie 2005; Cresswell 2006; Spinney 2006; Merriman 2007; Bissell 2007). My purpose is not just to add another form of mobility to the list, but rather to look at flying because it can throw certain issues into sharper relief, which might yield a richer understanding of embodied engagements with mobilities, with rhythm and arrhythmia, and with action and waiting.

Of course, flying has never been considered a mundane form of mobility but is widely perceived to involve high levels of skill combined with life-threatening risk, and, to balance those, has always required a great deal of specialized training. Henri Lefebvre explored the importance of such repetition-based training – what he referred to as 'dressage' (2004). In what he termed the 'military model' that repetition is extreme, 'pushed to the point of automatism and the memorization of gestures' (2004: 40). In other words, if rhythms are oft repeated, their practice may become automatic. As Tim Edensor (2009, no page numbers) has pointed out, 'rhythms become habitually and unreflexively embodied by those who participate in their performance, and are difficult to knowingly contravene.'

But training through repetition need not lead to automatism; it can also lead to practised and yet engaged responses to crises. In the case of a highly technical and profoundly embodied physical skill like flying, training – practice through repetition – must be relied on in an instant when the normal rhythms of flight are interrupted. To achieve the desired (even required) responses in the heat of the moment that practice, in pilot training, involves repeated simulation of crisis situations, with the hopes that such repetition will indeed make the desired response to arrhythmia automatic.

Women pilots of the 1920s and 1930s were articulate in describing such challenges when they occurred, and engaging those descriptions today can lead to a more nuanced understanding of rhythms, of 'dressage,' and of the importance of practised, affective, and embodied responses to rhythms and arrhythmias. By arrhythmia I mean specifically the changes in the relied-upon rhythms of aircraft engines, as well as changes in the predicted rhythms of the flights themselves; arrhythmia in this case can be mechanical or bodily – or both. Such arrhythmias in flight are significant to the pilots of course (often a case of life or death). But women pilots' rich descriptions of their responses to rhythms and arrhythmias in flight are significant also to scholars of mobilities for how they reveal the multisensual, affective, and embodied engagements that rhythms compel. As Lefebvre put it, 'No camera, no image or series of images can show these rhythms. It requires equally attentive eyes and ears, a head and a memory and a heart' (2004: 36) – as is revealed in these women's accounts. And further, while the experiences of women pilots throw issues of arrhythmia into high relief, all mobilities are suffused with rhythms and arrhythmias. As Edensor and Holloway point out in their research on a motor-coach tour in Ireland, 'Different rhythms possess capacities to affect different apprehensions and experiences: one drifts away from the unfolding spaces [outside the window] or finds oneself in a state of heightened attention to a particular event, sight or occurrence within our outside the coach' (2008: 499).

Arrhythmia in flight

In the case of early women pilots, arrhythmias in flight often demanded immediate action. Nichols, for example, detailed how she relied on her training when engine arrhythmia struck. Attempting, in 1931, to become the first woman to fly solo across the Atlantic, Nichols flew 'Akita,' her enclosed-cockpit Lockheed Vega with its strong 600 horse-power Pratt and Whitney supercharged 'Wasp' engine, from New York to a refueling stop at Saint Johns, New Brunswick on the way to her departure point in Newfoundland. Without enough fuel to reach an alternate refueling stop, Nichols was shocked at what she described as the 'tiny airport set like a bowl in the middle of surrounding wooded hills and cliffs .... How on earth could the heavy, fast Lockheed land there (1957: 155)?' With limited options, she set up for her landing and described what happened:

> [On approach] I slid in over the trees and edged through a narrow ravine. So far so good. Maybe my luck was holding. Dead ahead was the runway. I made an S turn for the proper approach and headed straight into the blinding rays of the setting sun. I couldn't look ahead to gauge the length of the runway, because ahead was the fiery glare. Only by staring straight down through the

cockpit window could I see even the edge of the runway [aircraft in the 1920s and '30s typically allowed no forward visibility once on the runway, so pilots were trained to take-off and land by using their peripheral vision to maintain an equal distance between the runway's sides. Nichols' case was more demanding as she could not see the runway ahead of her even before the wheels touched down and thus had little way of seeing where she was touching down, providing limited scope to judge the length of runway remaining to her: this was indeed coming in on a wing and a prayer].

In that split second, I realized [that] if I could touch the wheels on the ground before the intersection of the two runways, then I could clamp down on the brakes to stop the plane in time. … Cautiously, I eased back on the stick and kept my eyes glued to the edge of the runway. Now we were skimming the ground. Now the wheels must be about to touch.

Suddenly the dazzling blaze of the sun was doused by the shadow of a cliff and I saw to my horror that I had passed the intersection and still had flying speed. *Akita* was tearing straight toward an approaching cliff at eighty miles an hour. There wasn't a chance of stopping her in time to avoid destruction. Only one thing to do. I gave her the gun, pinning my last desperate hope on the chance that the speed of the ship at full throttle would pull her up and over the wooded crag [ahead].

You are not conscious of thought processes or even of fear at a time like this. There is only the awareness of stark necessity for lightning action. You act by conditioned reflex.

Akita shrieked at the suddenness of the climb, roaring toward the onrushing rocks. She lurched upward and I eased up on the stick to avoid dropping her into a power stall [which the fragile landing gear could not sustain]. The jagged edge of the cliff rushed at us. Here it was. I got set for the crash and it didn't come. By some miracle we had cleared the top of the crag by a hair. Even as I breathed a prayer of thanks, I saw another ridge ahead. Come on *Akita*! Good girl, *Akita*!

She struggled, quavered, made one last desperate effort—then came a splintering crack as the tail broke through the treetops. More rocks ahead—a deafening shuddering C-R-A-S-H—then paralyzing silence.…

Splintering pain and the silence of catastrophe.

My breath came in long, shuddering gasps. Automatically I moved my hand toward the instrument board, and noted with detachment that all the fingers worked. What should I have them do? Oh yes! Cut the switch. Always cut the switch when you crash.…Gas pouring down the sides from a broken fuel line.

> Boiling oil oozing out of the tank under my seat .... Get out before the fire starts
> ....
>
> Safety belt. Undo buckle. Hatch overhead. Push it open. Dear God how it hurts
> to move. But must ... Must get out. (1957: 155-157; last ellipsis in original)

Nichols broke five vertebrae and her airplane in the crash, and, the following year, Amelia Earhart became the first woman to fly the Atlantic solo. But Nichols had relied on her training, and the almost-instinctual responses that come from repetition, to get her through a balked landing and a nasty crash, as well as to escape the wrecked aircraft before a fire started. She had relied on her training in attempting to make a safe landing under difficult circumstances, and then, once that had failed, relied on her training again, despite her severe injuries, to rescue herself from the crash.

But, though crashes like Nichols' were fortunately not the norm, relying on training was. Nevertheless, each crisis demanded its own response. Lefebvre mentions in his brief evocation of automatized repetition that 'differences,' some 'unexpected' could occur in such rhythms (2004: 40). In flight, what is unexpected may be the crisis itself, but the difference in response to the crisis is often a result of the pilot's embodied and affective reactions.

Louise Thaden, who, in 1928 was preparing to attempt a women's altitude record, described flight-testing her Beech Travel Air (a comparatively light, open cockpit biplane) after it had been fitted with a powerful 'suped-up Hisso' (Hispano Suiza) motor just for the attempt:

> ... I took her up for the usual test hop. The engine ran smoothly with that hollow
> bellow which is such sweet music to a pilot's ears. I was circling over the airport
> [Oakland Municipal] at three thousand feet, checking the engine at wide open throttle,
> oil temperature, pressure, rpm's, when the engine began sputtering. It backfired,
> sputtered again, coughing, choked, and with final gasping wheeze—died.
>
> There was an awful stillness. The only sound was that of the eerie rush of air as
> it came against the wings and flying wires. (2004: 16)

At that moment, panic took over, but her training still guided her movements:

> Frantically pushing and pulling first one thing and then another [to try to restart
> the engine], all the while holding the nose down to keep up flying speed, I glanced
> over the side with the realization that objects on the ground were growing in size
> with startling rapidity. Banking the plane sharply, my heart must have stopped
> beating, for I felt suffocated. One slight error in judgment and the plane would
> be a twisted mass of useless wreckage .... (2004: 16-17)

Indeed, even with so little time, her practice and training helped her take charge of the situation:

> Tensely overcontrolling, I eased her down. 'Plan to overshoot,' I kept thinking over and over [—part of her training]. 'Then slip and fishtail …' [at the last minute to get the airplane down.]
>
> Thump! The wheels hit the ground. We rolled to a stop on my first [real] dead-stick landing. (2004: 17)

She had practised dead-stick landings again and again in training, but, as she learned, the reality was quite different from just practising, and her embodied experiences in the moment entailed more than just her training. Still, the repetition had paid off.

So, gaining confidence, and assured that the problem with the engine had been fixed, she tested the airplane again:

> I [flew] within gliding distance of Oakland airport for fifty minutes [to be safe]. The engine sang an even song. The airplane was perfectly rigged. It was one of those clear, crisp early December days that makes the blood tingle in your veins, gives a heady feeling of supremacy—a day good for the ego, one of those days when it's hard to fly right side up. So I flew across the bay to Mills Field, just to say 'Howdy' to the fellows there.
>
> Starting back to Oakland, I'd gotten well over the bay, settled back in the seat feeling satisfied with the goodness of life, when the engine began misfiring. I rode the ship down, and as I did I could imagine the cold, murky water closing over me. I felt I was in a horrible nightmare. Down, down—closer and closer to the strangling water. 'Shall I push the nose down and go in hard to get it over with, or settle in?'
>
> The engine coughed—my heart leaped—she picked up. Quickly I climbed. She coughed again, revving back. Sweat dripped from my hands. I was afraid, afraid to die. … I sat horror stricken, unable to think of anything to do, completely helpless from fright. The engine picked up again, intermittently starting and stopping; after about ten minutes, which seemed like hours, we landed safely at Oakland. [Shavings in the fuel tank had caused the problem.] (2004: 17-19).

So, though repetition is important in training, actually handling arrhythmia in the moment involves much more than mere repetition. It is an embodied and affective response, different each time it happens. And, though Thaden was able to rely on her training to get her through tough situations, the fears that seized her when her engine quit made her embodied and affective responses to crises, despite her training, different each time.

Rhythms of mobile waiting

Because dramatic moments of engine failures and crashes like those described above were rare, for most pilots, as for Nichols and Thaden, the rhythm of flying was more often about waiting: waiting alertly, for the presence or absence of arrhythmia. Waiting, as if in expectance, of arrhythmia because, as Edensor points out (forthcoming, no page numbers) 'rhythms continually change ... and where they are apparently repetitive and regular, they are apt to be punctured, disrupted or even curtailed by moments and periods of arrhythmy.' Early women pilots knew that, even when an engine was running well, that might change in an instant, and they were ever-watchful for signs of arrhythmia.

Howver, scholars of mobilities have tended to privilege movement and action over stasis and waiting, even though all forms of mobility can be mingled with practices of waiting – the body in motion may also be a body in waiting. As David Bissell has pointed out, 'the practice of waiting through spaces of mobility is an often-inevitable and frequent experience woven through the fabric of the mobile everyday ....' (2007: 277) – even the most highly mobile undertakings are co-mingled with stasis and waiting.

While Bissell conceptualizes waiting mainly in terms of processes such as sitting in airport lobbies, his point is that waiting itself is an active and embodied process, not always just something to be endured. Waiting itself, Bissell explains, is a particular kind of 'suspense' where 'the performance of waiting...heralds a heightened sensual attentiveness to the immediate spatiality' (2007: 285). And, importantly, as Bissell points out, despite popular understanding to the contrary, it 'does take *effort* and therefore some form of intentional action to wait' (ibid); waiting does not always simply happen, rather it must sometimes be wilfully engaged and endured.

Long flights present a good example of this, and Nichols described such experiences as she chased the women's distance record in 1931. Departing Oakland on the evening of October 23rd, she would not touch the ground again until the morning of the 25th in Louisville, Kentucky. Thus, for some 36 hours Nichols was in a state of intense waiting, alert at all times to the rhythm of the rebuilt Lockheed's Wasp motor. Climbing carefully to her cruising altitude, she wrote, 'I took a deep breath and settled into the rhythm of flight' (1957: 188). Lake Tahoe was her first landmark, 'shining in the moonlight' from a distance 'like a huge oval diamond in the blackness of the surrounding mountains.... I was filled with the familiar ecstasy of space and flight which were wings for my heart. I was soaring on top of the world, with the sky ... the moon ... and the stars as celestial company' (1957: 188-9).

But no sooner had she settled into the rhythm of enjoyable flight than

> The engine sputtered. [And] I snapped back to the reality of the smaller universe as I switched on another tank of gas figuring the fuel consumption.... Suppose my motor quit right now. I'd have to jump, steel corset, fractured back and all.

And then there would be the long, difficult trek to a railroad. But I had my compass and knapsack – I'd make it. Then I jerked my wandering thoughts up short. What nonsense! This good old Wasp engine wasn't going to let me down. (1957: 189)

Gradually, however, fatigue and boredom caught up with her and she reached for her coffee thermos. But the bottle rolled away and 'wedged [itself in] back of one rudder pedal' out of reach (1957: 190). Since in-flight emergencies, and landing itself, could require full rudder, 'This was serious', it was 'one of those stupid little accidents which could mean the difference between life and death' (1957: 190). She banked the airplane steeply until 'the thermos rolled far enough' out for her to grab it (1957: 190). That 'crisis,' she wrote,

> [k]ept me wide awake for a while, but [then] I began to feel drowsy. Everything was CAVU [ceiling and visibility unlimited—perfect]. I had plenty of altitude, unlimited visibility and in those days and at that late hour there was practically no possibility of meeting another plane. So I decided to rest my eyes for a few minutes by closing them. I knew I could remain alert even with my eyes shut for a brief time .... My ears were attuned to each throb of the engine, I'd know instantly if there were any change. After five minutes relaxation I was refreshed, and flew on for ... hours without incident. (1957: 191)

> ... Meanwhile I was dependent on the six hundred horsepower of a trustworthy engine, on the discipline of my own mind and body....

> As the hours flowed by, my eyes grew so-o-o heavy. [And by the second night,] it took maximum will power ... to force them open. All long-distance pilots have known the terrifying feeling of drowsing off and the tremendous effort it takes just to stay awake.

> ... But what was that? A streak of red across the darkness! At last the long, long night was over and I was witnessing the dawn of Ocotber 25. I had come safely through a night of indescribable splendor, *Akita* was still purring smooth and steady, and ahead [lay the end of the flight—nineteen hundred and fifty miles away from where it had begun and a new women's distance record later]. (1957: 192)

Without incident, the rhythm of flying was consumed by an intense, affective, embodied experience of waiting, a waiting that grew more intense and even painful as the flight time grew longer. And Nichols' waiting can be seen, as Bissell points out, as 'an achievement of a specific set of ongoing embodied tasks' that have their own rhythm (2007: 285) as she endured the passing hours of the long flight. Thus even highly mobile practices like flying are married with practices of waiting and stasis. But, as Bissell reminds, 'within every period of stasis, of stilling, is contained the potential to be otherwise, the possibility of rupture' (279) or arrhythmia. For Nichols, the sputtering of the engine and the loss of the thermos were not just interruptions in the rhythm of the flight but also in the long process of waiting for the flight to be over. Her experience shows how waiting itself is an intensely engaged, embodied experience, one that can be interrupted by arrhythmia at any time.

Conclusion

In recent years geographers have sought to understand experiences that are 'more-than-representational' and complexly engage affective registers (Lorimer 2006). To that end, some, like Spinney (2006), have sought to understand rhythms and mobilities through embodied practices involving their own participation – in his case through the experience of cycling on Mount Ventoux. Yet most historical work is based on archival traces – on representations of one form or another – and historical geography is most often precluded from the realm of the participatory because its events have already passed. Still, as Dewsbury, Harrison, Rose, and Wylie (2002: 438) have made clear, representations (such as archival traces) can be engaged 'not as a code to be broken or as an illusion to be dispelled' but instead 'apprehended as performative in themselves.' In this chapter, by engaging with the rich narratives left behind by women pilots of the 1920s and '30s, I have endeavored to do just that; to focus on the shaping of lives through 'everyday routines, fleeting encounters, embodied movements, precognitive triggers, practical skills, affective intensities, enduring urgencies, unexceptional interactions and sensuous dispositions (Lorimer, 2005: 84).' In particular, I have attempted to show, through these accounts, how rhythms of flying can be seen as 'architectures of sensation, narrative and embodiment' (Edensor and Holloway 2008: 499) that can reveal the affective responses to life-or-death circumstances of arrhtythmia, as well as the alert, engaged waiting that flying demands.

What the accounts left by early women pilots reveal are multiple ways in which the body-in-motion and the body-in-waiting engage rhythm and arrhythmia. For Nichols and Thaden, responding to arrhythmia relied on training, repetition, and practice, as well as on an active, affective, embodied doing-in-the moment that came about spontaneously through the intensity of the moment. Called 'dressage' by Lefebvre (2004), I have endeavored to show how such training-through-repetition leads not to automatized responses, but to affective and embodied engagements that are different each time they come about. When engine arrhythmia struck in flight, Nichols' and Thaden's training as pilots saw them through, even while their affective experiences in the moment yielded differences in their responses.

As I have also shown, a sustained engagement of rhythm and arrhythmia must rely on waiting – on the 'active doing of waiting' (Bissell 2007), with senses heightened to the body's immediate spatiality. Since engine failures and crashes were not the norm, the experience of flying – particularly on long flights

– was often about waiting, alertly, for arrhythmia. Waiting, always prepared (by training and practice), for what might be about to happen because, as Thaden put it, it wasn't always the case that the 'deep-throated roar of the engine became a melodious purr' (2004: 1).

PART IV
Dressage and Bodies

Chapter 12

Rumba and Rhythmic 'Natures' in Cuba

Shannon Hensley

Introduction

Rumba means so much. In the neighbourhoods, rumba meant a lot because earlier, people who practised rumba were discriminated against. There are people, you know, who study at university, people whose fathers have been to university. So, what is the difference between a dancer that learns at school and a dancer that learns in the street, a person of the street? The street dancer ... knows the rhythm better, the sound, because he was born with folkloric roots ... . And the dancer who learns at school? He has problems with the rhythm. This rhythm, you carry it in your blood, you're born with this—my father, my mother, an uncle, I don't know. My father is a percussionist and my mother is a dancer because they were born with this inside them. Aside from this, my grandfather and my grandmother were leaders of the Society for Negros. From a brother, an uncle, you take something. Because in the home is where you learn this, with them there beside you. You are there, eating it, hearing it, eating the rhythm, and without this no dancer, no dancer, could do it. (Interview with Dreiser, a rumba performer living in Havana, 25 December 2005. All names of research participants have been changed)

Hanging out with Dreiser in his shared second floor flat in *Centro Habana* on Christmas Day 2005, he shared with me photos and stories of what rumba rhythm and dance mean to him. During our conversation, he also demonstrated various styles of Cuban rumba: *yambu, guaguancó* and *la columbia.* In Cuba, rumba means, literally, to make a party. As a style of Cuban music and dance it differs significantly from the ballroom version of rumba, or rhumba, known widely throughout the world. On this day, Dreiser first demonstrated the couples' rumba dances *yambu* and *guaguancó*; the main difference between the two is tempo. Then, he showed off his impressive *la columbia* skills. *La columbia* is a competitive, acrobatic solo dance predominantly practiced by men although there have been, and are, famous female *la columbia* dancers as well.

I met Dreiser earlier that week at a rumba festival in Havana honouring the famous *rumbero* Chano Pozo, a famous musician, popular in the 1940s, who grew up in the *solares*, or overcrowded communal living arrangements in central Havana, widely known as the birthplaces of rumba. In the late 1940s, Pozo migrated to New York and played jazz with Dizzy Gillespie (Sublette 2004: 530-31). This rather

**Figure 12.1   Rumba band performing in *Los Sitios* neighbourhood, Havana,
            2005 (Photograph by author)**

under-publicized rumba festival ran from the 21st to the 28th December 2005, with
rumba performances every afternoon and evening. After seeing Dreiser perform in
the crowded and crumbling streets of Central Havana, I asked him for a 'dance
interview' so that I might learn about rumba and the embodiment of rhythm from
his perspective (see Figure 12.1). Dance interviews, largely a form of participant
observation, involve both dance and talk about dance (Hensley 2008, chapter 3).

During our conversation, as evident in the excerpt above, Dreiser suggests
that the way one learns rumba – where and from whom – affects the degree to
which rhythm becomes embodied. The spaces and relations of rumba training, he
argues, hinder or induce the transmission of rhythmic knowledge. He asserts that
the dancer or musician that learns rumba in the street, or at home, 'has' the rhythm
more so than those that are school taught. It is these understandings of rhythmic
natures that I examine further in this chapter, principally considering whether
notions of rhythmic embodiment deploy fixed, flexible or dualistic understandings
of the body. Throughout this chapter 'rhythm' refers to the temporal order of three
types of rumba music, *guaguancó, yambu* and *la columbia*.

In examining rhythmic natures, it also remains important to consider how rhythm
becomes associated with categories of bodily difference, such as 'race,' gender
and sexuality. For instance, in the excerpt above, Dreiser suggests that corporeal

knowledge of rumba rhythms must come from 'inside' the body, that the rhythms are 'in your blood' and, in some way, inherited or passed down through family relations. By referring to his grandparents' involvement in the *Sociedad de Negros*, Dreiser suggests that this history contributes to his ability as a rumba dancer as well. In this way, he connects his ideas about rhythm to family and 'race.'

What follows, therefore, considers understandings of rhythmic and racial natures in more detail to determine whether they invoke fixed or flexible understandings of the body. With reference to the work of feminist philosopher Elizabeth Grosz (1994) and social anthropologist Peter Wade (2002), I suggest that although 'race' remains a highly unnatural category of bodily difference, like rumba rhythms, racial identities become part of the developing body. First, however, I begin with a brief introduction to rumba and understandings of race in Cuba, which follows with a discussion about recent work in geography and critical dance theory around rhythm, music and dance. In this work, debates around the effects of dualistic thinking remain a central concern. The rest of the chapter considers how my research on rumba and rhythmic natures in Cuba can contribute to these debates.

Rumba and 'race' in Cuba

By some accounts, rumba emerged in nineteenth century Cuba in the port cities of Havana and Matanzas (Rodriguez, 1977; Daniel, 1995; Manuel, Bilby, and Largey, 1995; Sublette, 2004; Pérez, 2008). The majority of slaves brought into Cuba passed through these ports. Much later, as the process of abolition was underway, a number of freed 'blacks' found work in the ports. During this time of social change, including large-scale rural to urban migration, rumba grew from a relatively unknown type of music and dance into a widely popular genre, at least among the poor and working class (Blanco, 2000; Moore, 2006).

Either due to the prohibition of drums or out of material necessity, workers in the ports often improvised to create percussive instruments out of disused shipping boxes (Sublette, 2004). Many rumba performers continue to use wooden boxes to play *rumba de cajón* and rumba is widely understood as music that can be made from anything available: kitchen spoons, sticks, cow bells, kitchen tables and boxes (see Figure 12.2). As this suggests, rumba is generally considered music of the very poor.

There remains considerable debate over the 'origins' of rumba in Cuba however. Some argue that rumba is distinctly Cuban, and that it merges Spanish and African influences into a new musical form (Orovio, 1985; Blanco, 2000). Others assert, however, that rumba is a predominantly 'black' or Afro-Cuban practice, deriving most of its influences from African music and the descendents of slaves (Daniel, 1995; Basso, 2000; Pérez, 2008). Often, these sorts of assertion were made during interviews with rumba performers and audiences at rumba events. Generally, understandings of rumba as distinctly 'African' are widely held but less likely to appear in published works in Cuba (Moore, 2006). In many ways, this debate about

Figure 12.2   A rumba group plays on drums and wooden boxes at *el Gran Palenque* (the performance space of the National Folkloric Group of Cuba), Havana, 2006 (photograph by author)

origins resonates with contemporary concerns over how effective the policies of the Revolutionary government have been at eliminating racism against 'black' Cubans (Fuente, 2001a; Moore, 2006).

In the early 1960s, for example, many previously forbidden Afro-Cuban cultural practices, such as rumba, were heralded as symbols of the social changes enacted by the new revolutionary government (Kutzinski, 1993; Moore, 2006). The government also declared an end to racial segregation and social unity became an over-arching aim. To promote an ideology of resistance, change and a new equality the government often used Afro-Cuban music and literary practices as emblematic of these aims. Soon, however, debate on issues of racial discrimination was forbidden as it was seen as unnecessarily threatening to the project of building social unity. In this context, many Cuban music scholars argue that rumba represents *cubanidad*, or a collective sense of national identity wherein Spanish and African origins become merged, or unified, to create something uniquely Cuban. In this sense, notions of *cubanidad* aim to diminish the significance of distinct 'origins,' as well as ongoing processes of racial differentiation (Orovio 1985).

Regarding contemporary Cuba, scholars such as Pedro Perez Sarduy (1990) argue that claims of ongoing or renewed racial discrimination often deliberately ignore or dismiss the progress made by the Revolution. Concerns about Cuban social unity, however, often also work to heavily curtail public debate on issues of 'race' and racism (Moore, 2006). In this context, many rumba performers use ideas of blood, inheritance, ancestry and roots to assert the significance of distinct origins in their daily lives, including the practice of rumba music. Addressing understandings of blood, inheritance and rhythmic natures, therefore, allows us to learn more about how racial discourses work in people's everyday lives and whether references to nature deploy oppositional thinking about the body (Wade, 2002).

Rather than fixing the body, or opposing nature to culture, this chapter suggests that understandings of rhythmic embodiment suggestively blur the boundaries between culture and nature. Notions of rhythmic natures often refer to a process of becoming wherein culture and nature, or sociality and bodily materiality, are mutually constitutive, rather than fundamentally opposed. To consider rhythmic embodiment and issues of dualistic thinking further, I now turn to recent work on rhythm, dance and the body.

Rhythm, dance, and the body

An ongoing debate in both geography and critical dance theory regarding dance, embodiment and rhythm concerns the operation and effects of dualistic thinking. For instance, Jane Desmond suggests that dance research using an exclusively 'cultural studies' approach risks overemphasizing dance as text (1997a: 29-30). Here, Desmond criticizes research that approaches dance primarily as a form of cultural meaning production rather than an embodied experience, as well as work that relies heavily on textual forms of interpretation including Labanotation, a method of transcribing dance movement. Desmond claims that the emphasis on text and the omission of the material and expressive body in dance scholarship is largely due to dualistic thinking, especially the mind/body split. She argues that '(D)ance, as an embodied social practice and highly visual aesthetic form, powerfully melds considerations of materiality and representation together' (1997b: 2). Desmond stresses the need to consider the material and representational aspects of dance *mutually*, otherwise researchers risk 'the continuing rhetorical association of bodily expressivity with nondominant groups' (1997a: 30). She suggests, therefore, that approaches to dance, rhythmic expressiveness and embodiment need to consider the social significances of rhythm and dance – including how they represent social distinctions such as racial categories – as well as the materiality of the dancing body, while remaining committed to unsettling dualistic, determinist and essentialist accounts of the body (cf. Jacobs and Nash, 2003).

Non-representational theory also attempts to think dance as more than a 'text' or form of representation. Nigel Thrift (1997: 126) sets out the aim of non-representational theory in relation to the expressive embodiment of dance as 'attempting to provide a non-intentionalist account of the world.' He argues that the first goal of this non-intentionalist account is to 'provide a guide to a good part of the world which is currently all but invisible to workers in the social sciences and humanities, with their intellectualist bent, that part which is practical rather than cognitive' (ibid.). He goes on to claim that dance is about play: 'dance is not self-evidently about discourses of power and control. It is about *play*' (144). He defines play as 'gratuitous,' 'free,' 'not cumulative,' 'rule bound but with rules internal to the game,' 'difficult to command and control' and 'located' (146). Thrift's ideas about dance differ somewhat from Desmond's, however. She asserts that dance is a representation of social distinctions as well as a form of bodily expression and experience (1997b: 33). Thrift, however, suggests that dance is free and more about play than relations of power.

Other geographers also find merit in a non-representational approach to dance. George Revill (2004: 202), for example, points out that non-representational theory contributes to research because it shifts intellectual focus toward 'ordinary' practices and so 'provides opportunity for less hierarchic, more polyvocal studies of culture.' Tim Cresswell (2006: 73) claims that, '(T)his line of thinking has clearly opened up important new avenues for human geography.' I find some aims of non-representational theory compatible with this research as well, such as its attention to 'practices of subjectification' and 'the decentering of the subject' (Thrift, 1997: 127). I also agree with its attempt to pay attention to aspects of embodied experiences of rhythm and dance. However, as Catherine Nash points out, non-representational theory also seems to reassert dualistic thinking about body practices such as dance.

She notes that non-representational theory relies on a dualism between thought and action, the effects of which are four-fold: it is at odds with performativity because it ignores the productive role of power in constituting subjectivity; it lends itself to essentialist readings of dance as 'natural'; it 'does not provide a model for effective political strategy nor useful cultural politics' and its 'sense of dance beyond language appears unable, despite the stress on relational selves, adequately to combine a sense of the social or social relations with the unanalysable world of the precognitive or prereflective' (2000: 657) As Nash points out, by suggesting that the practical and the cognitive as well as play and power are necessarily opposed, Thrift asserts a clear division between bodily practices and thought.

These debates suggest that research on dance and the embodiment of rhythm is in a precarious, yet productive, position to contribute to understanding the relationship among performance, embodiment (materiality) and categories of bodily difference, such as race (Wade, 2005). These debates also suggest a need to remain aware of, work with and move beyond oppositional thinking. This chapter, therefore, attempts to move beyond dualisms by examining how bodily practices such as rhythmic responsiveness, *become unthought*, a tactic which suggests a rather different kind of relationship between thinking and doing.

Investigating understandings of, and techniques for, embodying rumba rhythms also allows us to learn more about the relationship between racial identities and the body. For instance, when pointing out the operation of dualistic thinking in relation to dance, Desmond (1997a: 41) notes that, '(I)n North America, it is no accident that both 'blacks' and 'Latins' are said to 'have rhythm'. This lumping together of 'race', 'national origin', and supposed genetic propensity for rhythmic movement rests on an implicit division between thinking and moving, mind and body. Of course, Desmond's concerns about the implications of dualistic thinking are well-founded; as she says, the body has long been associated with non-dominant groups through hierarchically valued binary pairs such as mind/body, human/animal, rational/emotional (Grosz, 1994; Price and Shildrick, 1999; Longhurst, 2001).

Desmond's point about the effects of dualistic thinking remains important when considering rumba rhythms as well. For instance, Louis A. Pérez, Jr. (2008: 199), points out that in the 1920s rumba was considered 'distinctly African and dismissed contemptuously by the Cuban middle class.' Moreover, rumba rhythms conjured ideas of 'sex and sensuousness, the lewd and libidinous... the primitive, the exotic, and the erotic' (ibid.). Clearly, these understandings of rumba deploy stereotypical and dualistic associations of blackness with the body and excessive sexuality. Although similar attitudes about rumba persist today, they are more openly contested through government endorsement of Afro-Cuban cultural centres and folklore programs in universities (Moore 2006: 176). At the same time, however, these arguments about the significance of rumba and the embodiment of rhythm remain limited. For instance, both Desmond and Perez fail to account for embodied experiences of rumba and rhythm by musicians and dancers. Attention to lived experiences allows us to question further how references to rhythmic natures work. Moreover, consideration of embodied experiences of rumba music will produce more possibilities for rhythmic subjectivity than those identified thus far (e.g. lewd, sexual, exotic).

Finally, while I recognize the importance of disrupting essential, categorical associations, such as those mentioned by Desmond, her claim that notions of 'having rhythm' rely on a body/mind divide fails to account for what 'having rhythm' might mean to different people in different places. To produce diverse understandings of rhythmic embodiment, we need to consider representations, materiality and embodied experiences in tandem rather than opposition (Cresswell, 2006; Edensor and Holloway, 2008). Therefore, it remains crucial to ask: how do rumba performers in Cuba experience rhythm and understand rhythmic embodiment? Does the notion of 'having rhythm' always rely on a body/mind, nature/culture divide? And, is it always a fixed, pre-determined propensity that is referred to when one claims that he/she/others 'have rhythm'?

As we heard from Dreiser, many rumba performers claim 'having rhythm' as part of their nature, yet they also often understand 'having' as a process of becoming with other people and through particular places (Wade, 2002). This suggests that the idea that one 'has rhythm' may not always be as essentialist a notion as Desmond suggests. At the same time, however, it is important to remember that the flexibility of racial discourses does not necessarily make them more open or inclusive (Nash, 2005). Therefore, to avoid dualistic thinking as well as essential associations of dance, rhythm and the body with certain groups of people, the next section emphasizes how rhythmic responsiveness is learned and *becomes* natural and instinctive.

Becoming natural: eating and sleeping rhythm

For all rumba performers, it is essential that rhythmic responsiveness becomes unthought, immediate and in a sense, natural as in 'second nature' (Wade, 2004). In fact, the fluid practice of music and dance demand the internalization of rhythm. In dance training classes, or hanging out, dancing at home, this often means that body movements are joined to the rhythm of the music through practices of counting out loud (Phelan, 1996). To make dance become play, or in performance, however, the practice of counting must disappear. The disappearance of the counting, thinking and learning that produces rumba dance creates the appearance of bodies seamlessly, naturally responsive to the time of the music. The process of learning the basic count of the rhythm is often intensive and is understood as getting the rhythm 'inside' the body, as one of my instructors suggests:

> Richard tells me that when we begin classes he will first have me practice playing the clave [a key rhythmic pattern that keeps time] with spoons, that way I will internalize the rhythm. He says that the dancer carries the clave inside and he points to his chest. 'That way, the dancer is in control, not the music,' he says. He then points out that once the dancer masters the clave, the drum can do whatever it wants, speed up, go crazy, and the dancer carries the clave, the time inside his body. (Fieldnote excerpt from dance lesson in Havana, 11 March 2006).

As this excerpt suggests, many rumba performers understand rhythmic responsiveness as a learned bodily skill. At the same time, that skill must become part of the body and unthought. Richard also indicates that the corporeal mastery of rhythm allows the dancer to stay in time with the music without necessarily matching the tempo. This is a very important skill for dancing rumba music as it is often played at a very quick tempo. As a dancer develops rhythmic skills, responsiveness requires less concentration and effort. This allows the dancer to focus on aspects of dance and music beyond counting, such as interacting spontaneously to the moves of a dance partner while staying in time. Corporeal mastery of rumba rhythms, therefore, helps the dancer avoid awkward, fumbling movements and generates the potential for playfulness, improvisation and innovation.

As Richard suggests, although many musicians claim that rumba is in their blood, they also remember periods of intensive training. During these periods, the 'blood' and body become invested, or entrained, with rumba rhythms. This suggests that ideas of 'having' rhythm in the blood also often involve process of 'getting' or 'placing' rumba in the blood and body. In this sense, 'blood' is understood as flexible, amenable to change and development. It seems, therefore, that references to 'blood' do not necessarily refer to predetermined genetic abilities. For instance, Yunier also recounts his experiences of eating and sleeping rumba, of getting the rhythm into his blood:

> Because I am not a schooled musician, but a musician from life, they say that I carry the music in my blood. Little by little, developing at first only on the clave. I first began playing the clave; afterwards I began with minor percussion. I then had two years developing my skills in a rumba group. I began to play the tresdos, and then the tresdos boxes. I realized, through prompting by the director of our music group, that if I didn't learn more, I couldn't continue in the group. So I took six months out from the group, in my house studying, and I hardly left the house, it was like I was eating and sleeping percussion. In that very same house my godfather, he is called Malanga, was a drummer. He was in the Cuban All-Stars and Yoruba Andabo. He taught me what rumba percussion was all about. (Interview with Yunier, a rumba performer in Havana, 29 December 2005)

To refer to the techniques of immersion that he uses, Yunier uses terms like 'eating and sleeping rumba' and he refers to important places and relationships: the manager of his music group, the house where he learns and his godfather, a famous Afro-Cuban musician, who also lived and played music in the same house. A *padrino,* or godfather relation, is often established through initiation into *Santería*, a religion derived from the practices of slaves known in Cuba as Yoruba or *lucumí.* Most rumba musicians also practice *Santería* and play music for *Santería* ceremonies. Although the padrino is not necessarily a 'blood' relation, he has enormous influence on his charge, as Yunier suggests. In a ceremony, the *padrino* receives and transmits the name of the saint, or *orisha,* given to his initiate by *Orula,* another *orisha.* These notions of 'inheritance,' cultural and spiritual transmission and descendence – deployed by Yunier later in this interview and by many rumba performers – are, I suggest, deliberately indistinct. They simultaneously invoke bodily, social and spiritual processes of becoming that are flexible and ambiguous rather than discrete and oppositional. These understandings of the body generally refer to the development of one's rhythmic abilities through relations to other people and places. They are not, therefore, entirely biologically pre-determined ideas of 'inheritance' or blood; instead they deploy a notion of bodily becoming that is both cultural and natural, as well as amenable to change.

Nosbelys, another rumba musician, also claims that rumba is in her blood and 'inside' her body. This, she says, is because she regularly practiced *bembé* (ritual drumming) at home with her extended family. Although she claims that rumba is in her blood, she also remembers a period of intensive rhythmic training:

> Shannon: And could you talk a little bit about learning the beat, the rhythm? Was there a time when you had to count out loud or something like this and now you don't have to?

Nosbelys: Oh, of course. At the beginning, whenever anyone learns whatever kind of instrument you always have to make the time, a count, a metric, so that you don't start off in the wrong time. So that you don't get ahead of the percussion – like when the clave goes one way and you go the other. And it took a little bit of work for me because I didn't know – I didn't have the knowledge for someone to say, 'one-two-three', and for me to instantly know when to start. After some time had passed, someone could tell me one, two, three, and I would know when to start. Then, without counting, I did it because I realised, then I knew where to start because so much time had passed, so many years since I had learnt how the rumba is. (Interview with Nosbelys, a rumba musician, 15/2/06)

As Nosbelys suggests, having rumba rhythms in one's blood does not necessarily refer to fixed or predetermined ideas of bodily abilities or rhythmic natures. She, and other musicians and dancers, still have to cultivate their skills, usually over a number of years. In this sense, rather than refer to a fixed idea of nature, blood invokes the amenability of the body to develop skills that become habitual, in some cases unthought, and in a sense natural. In these understandings of rhythmic natures, therefore, we find that both 'blood' and references to human and rhythmic 'natures' operate more flexibly than we might otherwise presume.

Moreover, when asked about the 'sexual nature' of rumba, Nosbelys suggests it is not 'sexual' at all. Instead, she argues that other dances such as tango, dances to romantic music, the salsa and the *són* are much more sexual because the couples dance with their bodies very close together whereas, in rumba, the couple spends most of their time apart (see Figure 12.3).

As Grosz (1994) suggests, however, it remains difficult for individuals, or even groups, to change the social significances of the body, body parts and bodily practices, including in dances like rumba. Equally important, however, is to note how bodies 'always exceed the frameworks which attempt to contain them' (Grosz, 1994: xi). Embodied experiences of rumba rhythms suggest important ways that dancing bodies exceed the existing frameworks for understanding rumba dance as, for instance, lewd (Pérez, 2008). For many performers, dancing and playing rumba is much more than a sexualizing, racializing and disempowering representational practice involving the body as a form of display (cf. hooks, 1997: 127). Rather, it is 'the power' of their life; they claim to experience healing, transcendence and a sense of expansiveness, rather than fixity or containment. Yunier, for instance, says that in rumba he feels 'completely alone ... inside himself and lost' (Interview, 29/12/05). Hector and Luis, also rumba musicians, claim that rumba is like medicine for them, it heals their bodies (Interviews, 2006). As these accounts indicate, responsiveness to rumba rhythms is both free *and* subject to command, immediate and learned, spontaneous and rehearsed, about play as well as relations of power.

Figure 12.3   The rumba dance *guaguancó*, 'Rumba Saturday' event, *el Gran Palenque*, Havana, 2006. In the *guaguancó*, the couple generally spends most of the dance separated by some distance (Photograph by author)

Conclusions

By addressing understandings of rhythmic natures through rumba in Cuba, this chapter has suggested that neither rhythmic abilities nor racial identities are fundamentally opposed to nature and the body. Instead, I have detailed how rhythmic responsiveness must become *part of* the body and 'second nature' as Wade (2004: 118-19) describes: 'we can see that the people learning and growing up in ... different environments over periods of many years in some sense develop different bodies, that is, bodies with different skills.' Susan Leigh Foster (1997: 241) also writes about this process specifically in relation to dance: '(E)ach [dance] technique creates a body that is unique in how it looks and what it can do.' She also notes how this process takes time: '(O)ver months and years of study, the training process repeatedly reconfigures the body' (ibid: 239). This chapter considered the techniques that rumba musicians and dancers use to 'join their bodies' to the different rhythms of rumba music (Phelan, 1996). I have described how the development of skills, such as responding to rhythm immediately, skillfully and innovatively (without 'thinking too much' about it) usually occurs over a long period of time and often involves periods of intensive training (cf. Hancock, 2005).

By examining techniques for learning to become 'naturally' responsive to rumba rhythms, I indicate how bodily materiality responds to and is affected by everyday habits and practices, including music and dance. As Grosz (1994: 191) argues, this ability of the body to be developed in a myriad of ways indicates that it is an 'open materiality,' plastic and pliable, open to augmentation and development, *not* pre-determined. As the body is developed along one trajectory, however, other possibilities are hindered.

As Grosz, Wade and Foster suggest, the development of bodily skills, whether understood as instinctive or learned, has significant effects on bodily materiality. Scholars such as Oliver Sacks (2007) have also pointed out that the body responds to music, not only on the conscious level of surfaces (by dancing, for instance), but also biologically and neurologically. In this sense, the idea that, through years of practice, some bodies acquire and 'have rhythm,' and others do not, becomes likely. Rhythmic responsiveness is a bodily skill that individuals develop over time in a variety of ways, some more conscious as processes of learning than others. Moreover, it is precisely because such abilities are developed rather than pre-determined, that not all bodies 'have' the same sorts of rhythmic skills.

How, with whom and where such skills are developed remains significant for their associations with racial identities, however. For instance, particular rhythms, such as rumba, have historical as well as ongoing associations with particular places and groups of people. Traditionally, as Dreiser indicates, rumba was practiced in predominantly poor and 'black' neighbourhoods: *solares* in Central Havana, the Pogolotti neighbourhood of Marianao on the outskirts of Havana, Guanabacoa and particular areas of Matanzas, places all known as sites of intensity for Afro-Cuban culture. The embodiment of rumba rhythms, therefore, often becomes the simultaneous embodiment of a racialized class identity.

As I have argued, however, that association is flexible and tenuous rather than fixed. So, although the association of rhythmic abilities and racial categories is neither genetically nor biologically predetermined, both bodily skills, as well as gender and racial identities, become part of the developing body and affect bodily materiality (Wade, 2004: 165; cf. Young, 1990; Grosz, 1994). To avoid social and biological determinism as well as dualistic thinking about the body, however, I have examined ideas about rhythmic embodiment with an explicit focus on techniques for developing the body and *learning* to become naturally responsive to rumba rhythms.

This chapter has discussed, for instance, how many people learn rumba in Cuba in ways that unsettle distinctions between play and training, natural and cultural, instinctive and learned, private and public. Many rumba performers learn to play music and dance at home, in the street, from family members and neighbours over many months and years. This type of training may lend itself more readily to naturalizing metaphors yet, as we have seen, the 'nature' referred to is often more flexible than fixed.

Therefore, if, as Grosz (1994: x) suggests, scientific ideas about bodies rely on 'everyday assumptions' as much as other forms of knowledge, *and* if both social scientific and biological ways of knowing 'are implicated in and in part are responsible for prevailing understandings of bodies' including dualistic notions; then, clearly, a diverse range of work addressing the 'nature' of the body is needed. One way of moving beyond dualistic thinking about the body in the social as well as biological sciences includes addressing how people understand their 'natures,' including their rhythmic natures. As Wade's (2004) work indicates, such investigations also allow us to learn more about how racial discourses and references to nature operate in people's everyday lives, and why, with many scientific declarations that racial categories do not exist genetically, the idea of 'race' persists.

Chapter 13

Equine Beats: Unique Rhythms (and Floating Harmony) of Horses and Riders

Rhys Evans and Alexandra Franklin

Introduction

Equine Beats are the rhythmical structures and practices which make up the experience of horse riding. Horse riding, or *equestrianism* is an increasingly popular pastime in the UK, a phenomenon we have explored elsewhere (Franklin and Evans, 2009), focusing on the ways equine activities facilitate the production of certain spaces in the rural landscape that we call Equine Landscapes.

Equine landscapes are socio-natures which provide the aesthetic, domestic, competitive, training and leisure spaces in which humans enact their relationships with equines. Unlike other 'agricultural' animals raised as part of food and commodity economies, equines provide humans with the opportunity to build partnerships which perform embodied acts in space. The walking, trotting, cantering, jumping and racing in these activities require a working partnership between human and non-human which, due to the size of the animal and the nature of the acts undertaken, cannot be solely coercive. Rather, both parties, different as they may be, act as one to produce acts neither would undertake on their own. The equine/human partnership forms an *actant* where the drives, needs and desires of both partners matter, and these frame and co-produce the distinctive socio-natures of equine landscapes.

In order to understand the cumulative impacts of the collective activities of horse riders, it is necessary to understand the ways the individual horse/rider experience is constructed. Previously, we have used an auto-ethnographic approach to begin this task. In this chapter, we focus on a single facet of the equestrian experience, using the lens of Lefebvre's *Rhythmanalysis* (2004) to illustrate how moments of rhythmic harmonisation form one of the main goals of horse riders and the epitome of the potential of an active unity of human and non-human. The liminal moments which this produces transform rural spaces of production into ludic spaces of motion and emotion. .

These moments are constructed out of both the rhythms of everyday life – which Lefebvre describes as subordinated to the '*measure time of work* which now is the 'time of everydayness' – and the 'great cosmic and vital rhythms: day and night, the months and the seasons, and still more precisely biological rhythms' (2004: 73). Further, the engagement in training, or *dressage,* provides a counter rhythm which, when properly applied, synchronises with those rhythms transforming the partnership from practice to performance. In so doing this transcends the limits of the *dressage* itself. Bodies disciplined in the nuances of stride, comportment and whole-body movement perform their rhythmical harmony in such a way that the training disappears, subsumed in the kinaesthetic union characterised by synchrony and synthesis. The performance is greater than the sum of its (training) parts, with transcendent moments composed out of the myriad rhythmical exercises which have been trained into their bodies. Between rider and horse, all these rhythms come into play and co-produce the composition 'Dressage horse/rider'.

There is a great diversity of equine experiences – as individual as each horse/rider combination. Together, these multiple rhythms concatenate to produce the overall equine landscapes. For this chapter, however, we focus upon one of Lefebvre's *garlands* of rhythms – that which produces the equestrian discipline of *Dressage.*

Dressage is both a discipline in its own right and one of the three sub-disciplines in the equestrian sport of *Eventing,* made up of *Dressage* (ground work, following a preset test), *Show Jumping* (a course of jumps constructed of coloured poles) and *Cross Country* (a course of solid 'rustic' fences, often incorporating 'natural' features such as hedges, ditches and water.

Dressage focuses upon the movement of the horse and rider over a complex course in a small arena measuring approximately 20 metres by 40 (or 60) metres. Commonly referred to as 'ground work', the movements consist of the horse completing figures on the ground employing specific steps, gaits and paces. We look at Dressage as a *composition* made up of different rhythms expressed across different time scales and in different activities. The elements in the composition of a horse/rider pair's experience of training, learning, practising and engaging in the discipline produce a rich and suggestive rhythmanalysis. Dressage represents one of the peaks of rhythmical expression in equestrian activities and riders often speak of a successful Dressage test as a special and specially rewarding experience.

Doing dressage

Dressage is often considered the most difficult of the three disciplines in eventing (MacLean, 2002; Newman n.d.). Julie Newman claims that, at its heart, its purpose is to 'enhance the natural movement of a horse' (n.d. 1). It requires horses to engage in 'many variations of forward, backward and sideways manoeuvres, and [has] very specific requirements of obedience, suppleness, gymnastic qualities and physical outline of the animal' (MacLean, 2002: 55). MacLean points to the origins of Dressage in cavalry, claiming that:

> Dressage tests obedience and suppleness over many prescribed movements. These evolved from the fast, agile manoeuvres required in war. As armory became an advantage in battle, horses were selectively bred to carry the extra

weight, and training requirements altered too. Movements became slower, more powerful as the emphasis changed from sheer speed to power: the horse had to carry most of its weight on its hindquarters, freeing its shoulders for turning relatively quickly. (99)

Reflecting this martial heritage, the tests in Dressage not only require a horse and rider to undertake a series of disciplined movements, but also stipulate a set of varied paces with which they are undertaken. Further, the transitions between paces (walk, trot, lengthening trot, canter, etc.) are carefully scrutinised and required to be undertaken smoothly and instantly when instructed.

There is also an aesthetics of Dressage in which these difficult movements and transitions are made to look easy and smooth. Thus, it is not only what is done but also how it looks which matters. Horses are meant to look 'light on their feet' –partially a product of the aforementioned power in the hindquarters, which lightens up the front legs and produces an appearance (and sensation) of 'floating' above the ground. High quality Dressage horses appear to move over the ground 'like a hovercraft' (55).

The discipline is regulated in extreme detail. British Eventing (the governing body) publishes details of every test in its *Rules and Member's Handbook* (2009). Dressage tests score the rider/horse on specific figures and on general competencies and comprise a series of movements which have to be 'performed' at specified places in the arena, with each movement attracting a mark out of ten. There are dozens of tests which range in difficulty from introductory to Olympic standard. In addition, marks are given for:

1. Paces (freedom and regularity).
2. Impulsion (desire to move forward, elasticity of the steps, suppleness of the back, and engagement of the hind quarters).
3. Submission (attention and confidence, harmony, lightness and ease of the movement; acceptance of the bridle and lightness of the forehand).
4. Position and seat of the rider, correct use of the aids.

There is thus an emphasis on ease and naturalness of the execution of movements that include instant transitions between walk, trot and canter, both up and down; movement along a diagonal whilst facing directly ahead; changing of the lead leg (left to right or vice versa) whist in movement; and tight circular movement. All movements are within the potential of the species, but it takes extensive training (and breeding) for any individual horse (and rider) to achieve some of them.

Dressage, however, means more than simply the skilful negotiation of a complex course. When asked, informants report that when everything goes right, and they and their horse are working together in concert, they achieve a special state of both heightened movement and heightened awareness. As one informant stated,

> When she [the mare] was swinging through her back, you feel like you've
> got complete weightlessness, you feel like you are floating over the ground. It
> only lasts 3 seconds or so at the beginning, but that's enough. That feeling, the
> difference in that feeling and the feeling you had before is massive. Before, you
> don't feel the power, when it's like that, you have it. You think they're going
> nice, but when they really go nice, wow!

This special feeling – weightlessness, floating, power – is one of the main reasons
given by many for taking part in the discipline, yet is rarely achieved, and only
after great investment in training and practice. While this 'floating harmony' is a
prime motivator behind these equestrian behaviours, the ultimate outcome and
reward of all the supporting activities, we also see it as a syncretic expression of
multiple rhythmicities.

Dressage, like the other two sub-disciplines of Eventing, is popular partly
because the point is to achieve this momentary unity of horse and rider through
following an historically developed system, expressed in the act of undertaking
the required activities in the competition. Some describe the liminal moment this
represents as a reward for all the training, the riding in all weathers, the early
mornings and the budgetary drain of owning a horse. Achieving this 'partnership'
with one's horse, in movement, is one of the main attractions of equine pursuits
– that union between human and non-human, engaged in an activity neither would
do separately but which both can do together. Interestingly, riders report that their
horses enjoy those moments too, that they also seem 'proud' or pleased with their
performance.

Horses communicate with their ears (position), eyes and the way they hold and
move their bodies. Most riders claim to 'understand' their horse and are comfortable
ascribing feelings (and often more) to this repertoire of equine communication,
interpreting them as displaying 'smugness', 'pride', and their opposites. How
much of this is anthropomorphising remains moot. It does, however, seem that
experiencing this floating harmony, so difficult to achieve, addresses a capacity for
pleasure in both human and non-human partners. Pleasure thus assumes the role
of driver and reward and is derived both from the investment in achieving it, and
the liminality inherent in the brief moment between two ground-bound modes of
existence. These Dressage activities and the satisfaction of these rhythmic drives,
calls and pulls in the particular places where they can be optimally satisfied, create
ludic spaces – equine landscapes.

Rhythm

Before exploring any further, the details of the horse/rider's world, we need to
focus upon our analytical lens. Lefebvre states that '(E)verywhere where there is
interaction between a place, a time, and an expenditure of energy, there is *rhythm*'
(2004: 15). By examining rhythms, then, we investigate the *interactions* between

horses, riders and their environment, the active constituents of the production of space (Lefebvre, 1991; Massey, 1995; Latour, 2005). Whether the actors are human, non-human, animate or inanimate, they are strongly implicated in the production of socio-natures (Macnaghten and Urry, 1998). Lefebvre's rhythmanalysis, therefore, provides a tool, a lens through which we explore one way in which equine landscapes are produced and maintained.

Lefebvre considers the components of rhythm to be perceptible in terms of *'repetition in time and space'* (2004: 6). He says that '(F)or there to be rhythm, there must be repetition in a movement.' (78), a repetition seen as *measures* in terms of speed, frequency and consistency. The musical terms of rhythm and percussion are used to describe this in terms of beats per measure – the equine world uses exactly such terms, with trot measured as two beats per measure, canter as three and gallop as four. Some rhythms are slow, measured in days, weeks and seasons; others refer to corporeal rhythms of the bodies of the horse or rider. Some rhythms are marked by presence and others by absence. The medium through which rhythms are expressed varies from the corporeal (hungers, hormones) to the ephemeral (aspirations, competition). They come together in the *riding* – the ultimate expression of equine rhythms in a partnership between horse and rider.

Lefebvre uses a number of key terms including *polyrhythmia, eurhythmia, and isorhythmia* (67). *Eurhythmia* describes a polyrhythmic relationship between rhythms (and ensembles of rhythms) which has a harmonic resonance. As Lefebvre describes it, '(E)urhythmia presupposes the association of different rhythms' (67). Each living body is an association of different rhythms – the rhythms of the organic body with its homeostatic mechanisms, drives, responses and preferences. Lefebvre refers to a 'eu-rhythmic body, composed of diverse rhythms – each organ, each function, having its own – [which] keeps them in *metastable* equilibrium' (20). This quality of *eurhythmia*, or a harmony composed of many independently-existing different rhythms, is an important quality he associates with a 'naturalness' constructed out of reference to the biological organism:'(T)he theory of rhythms is founded on the experience and knowledge [*connaissance*] of the body' (67). For Lefebvre then, a polyrhythmia usually lies beneath the awareness of the organism, *producing* the single entity which is the individual horse or rider, which in this sense, may be eurhythmic wholes.

The same concept appears in the field of neurobiology, where rhythm is considered an integral part of the neurological organisation of animal life. Merker, Madison and Ekerdal (2009: 4) state that 'rhythmic movement is a commonplace in the animal kingdom, being essential for efficient locomotion in water, through the air, and on land (von Holst, 1973). These and other rhythmicities such as those of respiration are generated endogenously at quite basic levels of neural organisation'. These *endogenous* rhythms range, therefore, from those which usually (but not always) operate at a level below perception such as heart rate, through to locomotion which is more susceptible to conscious influence.

Further, they claim that the endogenous rhythmicities of individuals can be entrained into an externally applied regular rhythm. Having established the internal eurhythmia of animals as a kind of autopoietic or emergent property of their neural organisation, they claim that there is a similar phenomenon which allows *multiple* animals to engage in rhythmically harmonised movement, giving examples ranging from frogs croaking in concert to humans drumming or tapping their feet. This effect is called *isochrony* and is defined as 'capacity to "entrain" to an evenly paced stimulus'. *Entrainment* is a term used in neurology and psychoacoustics to describe the way the rate of brain waves, breaths or heartbeats vary from one speed to another through exposure to an external rhythm. Entrainment is thus the mechanism by which two independent organisms internally synchronise with each other rhythmically and isochrony is the pulse to which they entrain and the resulting rhythmic harmony. The pulse used to effect this isochrony is also called the *tactus*, a term used to describe the pulse which underlies the 'temporal structure of all rhythmic music' (Arom, 1991: 179).

Within neurobiology, therefore, we have eurhythmic bodies which are produced in part from their endogenous rhythms. Their rhythms are synchronised not only as an act of conscious intent but, through entrainment, as a subconscious effect of neurological organisation. This produces measurable changes in the organism as evidenced by entrained breathing or locomotor activity. This application of isochrony through the use of a tactus is one way in which Lefebvre's isorhythmia – that is, the imposition of an 'equality of rhythms' (2004: 87) – is expressed, and in the case discussed here, between horses and people.

The process of training horse and rider to negotiate the physical challenge of a Dressage test together is a process of harmonising the rhythms of these eurhythmic wholes. By 'equalising' their rhythms the horse and rider achieve a isorhythmia which allows them to engage in something neither would do individually, something which both can find satisfying. These ensembles of rhythm interact together producing the isorhythmic whole which is the epitome of their mutual entrainment – the moments of floating harmony which take them outside a ground-bound existence.

Motion and emotion

Before exploring the rhythmicities of horse riding further, we point to the motivations for riders and horses to sustain the often tedious process of this repetitive, rhythmical training. Key to this is the *feelings* of both participants. As one rider stated, 'the feeling of being 'at one with the world again' which I carry after riding makes the personal inconvenience and costs seem minor'. For her, it is the emotions which provide both the motivation and the pay-off for submitting herself and her horse to the repetitive rounds of training, and her time and budget to sustaining the broader cycles which are a core component of their health and ability to perform. Further, as the earlier quote reveals, her response to the sensation of floating, of the disappearance of the rhythms in their momentary conjunction is one of pleasure, excitement and strong emotion.

Horses too are capable of emotions and feelings. Emotions do not require reasoning and as Andrew McLean asserts, '(E)quine researchers agree that any higher mental processing abilities in the horse are, if present at all, poorly developed.' He goes on to assert, 'When a horse behaves in ways that don't suit us it is wrong to say "He knows what he did wrong" or "He understands". There is no understanding in the horse – he simply reacts to situations, events, aids etc' (op. cit: 4). In this case, 'reactions' are feelings. Feelings are important to beings which do not have the ability to reason or verbalise: they are key to learning because positive reinforcement, by definition, *feels* good. Without feeling pleasure and aversion, the system of learning which has made the horse survive and adapt as a species would not work.

Sheller, speaking of the embodied pleasures of automobility refers to a similar link between feelings and movement. She states, '(E)motions, in this view, are a way of sorting the sensations of the non-cognitive realm, which occur through the conduct and movement of the body .... Motion and emotion, we could say, are kinaesthetically intertwined' (2004: 2). For the horse, feelings are important and acute principally because they do not have alternate media for cognition. Proprioception and kinaesthetic feeling are paired with emotions even in organisms which can verbalise. Feelings are part of the feedback mechanism by which an organism monitors its locomotion. Movement is thus paired with emotion.

Humans too associate pleasurable emotions such as exhilaration with a wide range of bodily movements. Motion and emotion are linked recursively. In some cases, motion can entrain emotion (a 'runner's high' or various types of dancing) and in others emotion can entrain motion (to 'jump for joy' or 'it makes me feel like dancing!').

Referring to automotive complexes, Nigel Thrift claims there is a need to,

> … understand driving (and passengering) as both profoundly embodied and sensuous experiences, though of a particular kind, which "requires and occasions a metaphysical merger, an intertwining of the identities of the driver and car that generates a distinctive ontology in the form of a personthing, a humanized car or, alternatively, an automobilized person" (Katz, 2000: 33) in which the identity of person and car kinaesthetically intertwine. (2004: 46-7)

In a similar way, the rider and horse form a kinaesthetic unit when performing activities such as jumping or Dressage. They too represent a kind of *personthing*, only in this case, a *horsepersonthing*. The difference is that the horse, unlike a car, has an independence of identity and a will which is capable both of frustrating a human's attempts at building a partnership and of participating in the creation of precisely such a partnership – one which often stands on a liminal margin where the achievement of this transcendence creates an entity which is capable of more than is either of its two parts.

This kinaesthetic assembly of motion and emotion is constituted out of rhythms – the eurhythmic bodies of the partners; the rhythmic training of which they partake, and, as we will see, the isorhythmic union they achieve when, through the practice of their embodied rhythmic strategies, they achieve the syncretic *leap* which takes them beyond their individual selves.

'Dressage', or disciplining bodies with rhythm

A discussion of the equine world through the lens of rhythmanalysis cannot avoid Lefebvre's use of the term *dressage*. Indeed the two terms – dressage and Dressage form the heart of this chapter. We use the term 'dressage' to refer to the French word *le dressage* as used by Lefebvre, and 'Dressage' to refer to the equestrian discipline.

For Lefebvre however, '(O)ne can and one must distinguish between education, learning and dressage or training (*le dressage*).' (2004: 20) Here then, it is the training which is of interest, not the performance. This dressage involves both the horse and the rider. Indeed, Lefebvre's focus is primarily human. 'Humans break themselves in (*se dressent*) like animals. They learn to hold themselves. Dressage can go a long way: as far as breathing, movements, sex. It bases itself on *repetition*' (ibid.). Repetition lies at the heart of the training undertaken by riders and their horses and the rhythms of training structure their lives.

This is training of a very embodied sort, for both horse and rider. Learning to ride requires three separate categories of learning tasks, of 'training'. One involves corporeal and kinaesthetic learning – positioning and moving the body in specific ways in response to movement and positioning of the horse. Similar types of kinaesthetic learning occur in skiing, surfing and other human sports. The bodies are trained to hold, move and respond in specific ways which enable the performance of the desired athletic feats. A second type of learning involves conceptual learning – learning to recognise vocabulary, features, factors and situations, and respond to them. It involves matters of strategy and tactics. Strategically, trained bodies aim to accomplish the desired object, and tactically, gain a knowledge of the intricacies of training which allows for self-reflection and improvement. The third type of learning involves communication – learning a language. This language is mostly physical and embodied in nature, built out of the same physical contact with the horse as kinaesthetic learning. It involves a subtle language of touch, position, response and engagement and is the foundational basis of the 'partnership' between horse and rider. Horses have a remarkable facility to interpret this language of touch, based upon the seating of the rider's body, receiving commands, feelings, moods and emotion such as tension or confidence. As a result, riders have to learn that every part of their bodies which touch the horse communicates *something* and learn to control that communication. Like learning any language, this too requires repetition and iteration.

All these types of learning require *iterative learning* – repeated iterations of small physical, mental and communication tasks which combine so that horse and rider do exactly the right things at the right times to perform, for example, a 'lengthened trot' – one of the movements which can produce the 'floating harmony'.

This iterative learning is emphasised formally within the discipline too. Lessons for ground work are distinct from and often precede training for the jump. The focus of the *dressage* or training is on both horse and rider, with instructions to the rider directed at training the responses of both. Training is based upon behavioural ethology – positive and negative conditioning, incremental modelling of behaviour, and the building of neurological pathways which produce new behaviours (AEBC, 2007; Maclean, 2003). Thus it is repetitive, iterative and progressive. It has a rhythm and a trajectory and a pace which is based upon the two biological organisms involved. Both humans and horses learn this way. Lefebvre states that dressage is:

> ... able to bring about unity by combining the linear and the cyclical. By alternating innovations and repetitions. A linear series of imperatives and gestures repeats itself cyclically. These are the phases of dressage. The linear series have a beginning (often marked by a signal) and an end: the resumptions of the cycle (*reprises cycliques*) depend less on a sign or a signal than on the general organisations of time. (2004: 39)

In this case the 'unity' is the production of the floating harmony previously mentioned. At multiple levels (horse, rider, horse and rider) training involves the repetition of simple acts until they become conditioned behaviours. A new behaviour -which is generally an increment of an already-learned behaviour – is introduced and repeated again and again until it too becomes consolidated. As Lefebvre suggests, this assumes a rhythm, a series of repetitions and innovations which give it a scope or trajectory, a movement towards a goal which is approached in incremental steps. Signalling the rhythmic nature of this process, Lefebvre asserts that '(T)he *sciences* of dressage take account of many aspects and elements: duration, harshness, punishments and rewards. Thus rhythms compose themselves' (40). These are the components of rhythm and are the day-to-day constituents of the rhythm of training, or *le dressage* of riders and horses within the discipline of Dressage and other equine disciplines.

Within the boundaries of dressage operate another *garland* of rhythms – those of the bodies of the horse and rider. The endogenous rhythms of the heart, breathing and locomotion are trained to strengthen them. Here,

> rhythm appears as regulated time, governed by rational laws, but in contact with what is least rational in human being: the lived, the carnal, the body. Rational, numerical, quantitative and qualitative rhythms super-impose themselves on the multiple *natural* rhythms of the body (respiration, the heart, hunger and thirst, etc), though not without changing them (9).

The point of applying this rhythmical training is to 'change them' – to cultivate the potentialities of the organisms, individually and collectively, to achieve the desired performance. This combination of the linear and the cyclical affords both pattern and innovation or novelty. Without novelty there is no learning and without pattern there is nothing to learn. The cyclical and the linear components of rhythm are key to learning, progress and, in the case of Dressage, performance. Training for fitness involves the repetition of (cyclical) acts which involve, for example, aerobic strength, with incremental (linear) increases through the introduction of new tasks, new challenges. The apparent 'ease' of the floating harmony masks a need for extreme strength and fitness. Careful management of both the horse and the rider's fitness is needed to both avoid injury and arrive at a state where there is sufficient strength (and skill) to perform the tests.

Further, the ability to perform is affected by the rhythms of the lives of the horse and rider outside of these trajectories of training. Lefebvre refers to this: 'everyday life remains shot through and traversed by great cosmic and vital rhythms: day and night, the months and the seasons, and still more precisely biological rhythms' (73). For the horse, for example, the biology of their periodic sexuality affects a mare's body and a stallion's attention span. The seasons too can affect their energy and attention levels through varying cycles of light/dark and the nutritional content of the grass they eat. For the rider, the cycles of their life outside equitation (work, family, commuting, budget) affect their fitness to achieve the union of rhythms which is embodied in those floating moments. Thus, rhythm conspires against performance as well as supports it.

Rhythm thus is not only an aid towards harmony. Like tones, they also can be disharmonious. Lefebvre terms this *arrhythmia*, '[the] all becoming irregular (*dereglement*)... of rhythms produces antagonistic effects. It *throws out of order* and disrupts' (44). The effect is the breaking of a state of eurhythmia or the prevention of its attainment. 'In arrhythmia, rhythms break apart, alter and bypass *synchronisation* ...' (67). The aspiration to achieve a state of eurhythmia between horse and rider is therefore mitigated by the interruptions of other rhythms and cycles in both their lives. Yet dressage or training may overcome these disturbing influences, their strength evidenced by the fact that occasional attainment of the ultimate unity of mutual rhythms can be achieved despite the intervention of arrhythmic influences. As Lefebvre says, 'dressage does not disappear. It *determines the majority of rhythms*' (40). The legacy of training, therefore, is to allow the pair to recover the eurhythmic moment despite the distractions of arrhythmic dis-harmonies.

There are thus multiple rhythms to this learning. There are the endogenous rhythms of the two different bodies. There is the formal practice of training or *dressage* with its iterated rhythms of learning. And there are a larger set of rhythms and cycles which operate at a diurnal, weekly or seasonal periodicity. The rhythms of dressage are also made up of early morning training outside in a 'school' (a fenced flat surface, often modified to provide cushioning and grip, which mimics a dressage or show jumping arena); weekly 'lessons' with expert teachers; daily or frequent 'hacking' and physical conditioning, and weekly participation in competitions – these form the rhythmical boundaries of Dressage and form the basis for their achievement of the eurythmic and transcendent moment of 'floating harmony'.

Isorhythmia – achieving unity between horse and rider

In discussing dressage we have focused upon the training of the bodies of horse rider in ways which make each capable of achieving moments of floating harmony. There is one further rhythmic strategy which is constitutive of the creation of the floating moment, and which is overtly practised in the equine disciplines. This is the uniting of the two rhythmic bodies in such a way that they create this kinaesthetic act of transcendence. The direct object of this repetitive training is to teach both bodies to engage in positional and balance postures which complement each other in the attainment of the floating stride. This is expressed in the judging of Dressage as 'the position and seat of the rider' (BE, 2009). Given that a horse has been trained to achieve a stride, the rider is trained to synchronise with this, to 'not get in the way of the horse'. Yet at the same time, the rider retains a command function in that the pair's objectives are human-directed. As the rider is simultaneously training the horse and training their own body, they are also immersing themselves in the demands of the test. This imposes rhythmical disciplines on them both, in that the size and number of strides required for each manoeuvre is set by the creation of the test which, at the completion of the test, measures conformity to these constraints. Thus the rider imposes another rhythm – that of the human-designed course – on them both. If rhythm is measured by 'a differentiated time, a qualified duration ... of repetitions, ruptures and resumptions' (Lefebvre, 2004: 78), so too is the performance of the horse and rider in a Dressage test. The imposition of this exogenous rhythm is accomplished through the creation and application of a tactus (the beat or pulse which harmonises their individual eurhythmias) by the rider. Composed out of the rhythms of the course – the time limits which determine pace, and the distances which determine strides – the tactus is designed to ensure they are at the right place at the right time in the test, that the horse is at the right part of its stride to accomplish the manoeuvre, and that the rider is in the right place in their seat to synchronise with and support this. This is the mechanism of the attainment of a state of isorhythmia, which in this field is the ultimate outcome of the training. The set of exogenous beats is first internalised by the rider and then communicated to the horse. Through training, the horse learns to recognise these signals and adopts the correct stride, pace and comportment. Thus, 'intervention through rhythm ... has a goal, an objective: to strengthen or re-establish eurhythmia' (68). Or in this case, isorhythmia.

These multiple cycles of learning, training and entrainment are repeated, practised and fine-tuned over and over again in iterative aspirations of rhythmic unity. Each attempt offers the potential to differentiate in one small beat, one sudden syncopation, a transcendence of the very process which switches from training to performance and which makes the united pair *float* just a bit smoother.

Conclusion

This rhythmanalytical study of the equestrian discipline of Dressage illustrates the fecundity and potential of Lefebvre's conceptualisation to constitute one small aspect of lives-in-a-landscape, a mechanism for the production of space. It offers an opportunity to see how '*Rhythm* reunites *quantitative* aspects and elements, which mark time and distinguish moments in it – and *qualitative* aspects and elements, which link them together …' (9). The practice of *dressage* or training, rhythmic in itself, links emotions and motions and trains to produce a synthesis of the two, and this dialectic of horse, rider and horse/rider produce the liminal experiences of 'freedom', 'floating' and 'not touching the ground'. These experiences represent the epitome towards which many riders aspire.

The constitutive rhythms in this rhythmic composition are both endogenous and exogenous. The individual bodies carry their own time signatures built of the multiple processes of different rhythmical periodicities which keep them alive, and when healthy, exhibit a state of eurhythmia. Their individual locomotor frequencies are similarly different but capable of polyrhythmic harmony. The practice of embodied learning, or dressage itself, consists of the application of linear sequences of cyclical practice which both constitute and mark progression towards a physical goal in terms of locomotion performance. Not only is time marked, but physical progress is also encompassed by these rhythms.

Key to this analysis is the *union* or synchronisation of the eurhythmic bodies of horse and rider. One of the unique features of equine pursuits is the challenge of forming a partnership between human and non-human, especially given the size and power of the non-human partner. Union is not something which can be enforced. Rather, this partnership is formed through repetition, through the rhythmic application of embodied incremental training, the ultimate purpose of which is to produce what in the equine world is called 'partnership' and 'trust', via a state of isorhythmia, where an external rhythm is imposed to harness multiple eurhythmic bodies in unison, enabling performances which are syncretic and capable of producing effects which transcend the capabilities of either partner.

In creating this union, the application of these rhythmical structures supports responses which are non-physical. The elation of the rider in their moment of 'floating harmony' is claimed to be echoed in different-but-similar behaviours of the horse. In this situation, the connections between the physical and the psychic are equally empowered by the application of dressage. Rhythm provides the link and mechanism between them. The transcendent moment between ground-bound worlds, of 'floating harmony', is dependent upon the effects of all the rhythms and cycles which have gone before it. These constitute that moment, which exists almost as a standing wave constructed out of all the component rhythmicities which have produced it.

Equine landscapes are distinctive spaces produced by equestrian activities. As more people take up these activities, their regular, repetitive presence moving through the British countryside will influence the creation of new distinctive landscapes. The potential of these activities, however, to influence the production of new spaces of leisure, sport and pleasure is directly dependent upon the strength of individual commitment to repetitive embodied activities in spaces which are well suited to them. This rhythmanalysis of the production of equestrian bodies, human and non-human, through isorhythmic interventions which take place in a landscape provides a productive lens with which to view one of the drivers of this socio-nature at the scale of individual horse/rider combinations. The strength of the pull of motion/emotion in equestrianism mirrors similar strong attachments in cultures of automobility where, as Sheller claims, 'such 'automotive emotions' – the embodied dispositions of car-users and the visceral and other feelings associated with car-use – are 'central to understanding the stubborn persistence of car-based cultures' (2004: 4). Herein lies the potential strength of equine pursuits as an agent of change in the British Countryside. We have looked only at micro-scale rhythmicities of equestrianism. Scaled up to larger levels, the eurhythmias produced by collective equine activities has similar potential to powerfully affect the production of 'nature' in the countrysides in which they are expressed.

PART V
Rhythms and Socio-Natures

Chapter 14

'The Breath of the Moon': The Rhythmic and Affective Time-spaces of UK Tides

Owain Jones

But the most admirable thing of all is the union of the ocean with the orbit of the moon. At every rising and every setting of the moon the sea violently covers the coast far and wide, sending forth its surge, which the Greeks call *reuma*; and once this same surge has been drawn back it lays the beaches bare, and simultaneously mixes the pure outpourings of the rivers with an abundance of brine, and swells them with its waves. As the moon passes by without delay, the sea recedes and leaves the outpourings in their original state of purity and their original quantity. It is as though it is unwittingly drawn up by some breathings of the moon, and then returns to its normal level when this same influence ceases. (*Opera de Temporibus*, Section XXIX, the Venerable Bede, 703 AD)

Introduction

I have been thinking of buying a tide clock, a clock that does not tell you the time, but, instead, the state of the tide at any given moment. They can be bought in marine chandlers and are of use to those who work, or 'play' with the sea and its margins: a shipping pilot who guides large ships into tidal ports at high tide; a farmer who has stock grazing on intertidal pastures which may be at risk of drowning at the highest tides; someone who sails recreationally out of a tidal harbour; a collector of shellfish for a living when the tide is low; an ornithologist who watches (or cares for) the birds that feed on exposed intertidal land; or someone who fishes from a seawall. The clock will show when the tide is rising, when it is at its highest, when it is falling again, when the water level is lowest, and then turning to rise again – starting the cycle over. This tide clock may be as important to these people as their chronographical timepieces which tell the hours and minutes of the day.

Much ecological, social and economic life has circadian rhythms, driven by the daily rotation of the earth in relationship to the sun. Although much has been written about the extent to which social, human (clock) time has broken free from this profound natural rhythm, it still remains a ubiquitous pattern of life – not least through human and non-human body clocks which are finely tuned to the turn of day and night and result in patterns of sleep and wakefulness and much besides (Foster and Kreitzman, 2004). Along the coast of the UK and other places of high tidal fluctuations, tidal rhythms also influence the temporal patterns and rhythms of life. Thus it could be argued that forms of lunisolar, *hybrid* temporality occur in these places – driven by the interlocking rhythms of day-night (solar rhythm) and tidal rise and fall (lunar rhythm).

I offer a personal example. My father was a farmer whose working day was generally diurnally and seasonally dictated – getting up at first light (or before in the winter) to milk; finishing off work when evening came. But there was another pattern woven through this. We grazed sheep and cattle on intertidal grazing land near the docks of Cardiff, Wales, a few hundred acres of flat land where the protective sea walls snaked inland leaving the grazing open to the higher tides. At the times of monthly, and especially seasonal high tides, my father, often with us children along as helpers, would ensure the stock were safely on, or behind the seawall. In the organised chaos that was my father's desk – the farm office – a small yellow *Arrows Tide Table* for the Bristol Channel was always in its place (his equivalent to the tide clock). (Jones, 2005).

This chapter explores the temporal rhythms of tides and their effects on everyday life. Intertidal spaces and their land margins are spaces of extraordinary richness for entangled human and non-human becomings. Economy, ecology, culture, identity, all meet transformative moments at the water's edge, yet with(in) the tides that edge is in a perpetual, complex rhythm of ebb and flood.

I briefly illustrate material cultures of tidal rhythms before considering intertidal areas and their margins as *affective spaces,* drawing upon examples from art and literature. I have long witnessed the tidal ebb and flow of the Severn estuary so I bring that personal experience to bear. I have collected a range of references to tidal culture in British literature and art, for traces of these 'other-timed' landscapes and the lives within them can be found in these representations of landscape, and are also present in a whole host of practices and accounts of local landscapes and cultures. Thus I follow a tradition within human geography of using literature and the arts as sources of material to consider landscape, relationships between nature-culture, sense of place and identity (Cloke, Philo and Sadler, 1991).

There are scientific studies of tides (Cartwright, 2000) and work on tides and ecology (Carson, 1961) yet while tidal processes are some of the most obvious, ubiquitous and powerful natural processes/rhythms in costal landscapes, they have attracted scant attention in cultural geography, landscape studies or time geographies.

Tidal patterns, rhythms and spaces

If I choose to buy a tide clock there is a complication – to *where* should I set it? You not only have to set the tide time – find out an exact time of high or low water and set the (single) hand to match – you have to decide which point on the coast you will set the clock to. Tides are generally semi-diurnal, rising and falling roughly twice a day. But there is much variation, and the precise timings and rhythms vary from place to place. The moment and point of high tide moves around the globe daily, following the orbiting moon, washing along, or piling up against, differently orientated coastlines. Tidal processes, ranges and timings can vary markedly within a few miles of coastline.

The complexities of tidal rhythms are created by the relational motion of earth, sun and moon. Briefly, it is the gravitational pull of the moon that draws the mass of the oceans towards it as passes overhead. The more distant sun, whose gravitational influence is much less, either dampens this when pulling against the moon, or heightens it when they pull in concert. Thus tides not only generally rise and fall daily, but have monthly, seasonal and yearly variations which correspond to the complex *pas de trois* of these three heavenly bodies (McCully, 2007).

The basic tidal sequence is a continuing cycle of *low water*; *the flood* (tide rising), *high water*, *the ebb* (tide falling), and *low water* again. The all important *turn of the tide* marks the moment when the cycle switches from fall to rise, or rise to fall, and at the turn of the tide there is a brief period of *slack water*. The highest tides, which occur in conjunction to the equinoxes, are *spring tides*, and the lowest, *neap tides*. Importantly, the spring tides not only mark when the water level rises highest, but also when it recedes to its lowest. In other words, spring tides consist of the largest range between high and low water, whereas neap tides have a much smaller range between high and low water with correspondingly less intertidal area exposed/covered. At the spring tides a landscape such as the Severn estuary changes very dramatically in the space of six or so hours between low and high water. At the neaps much less happens but the cycle is still in place.

This base rhythm of the tides is affected by all manner of variation. Firstly coastal location, form and orientation all affect local tidal ranges. The oceans of the globe have differing tidal characteristics (Carson, 1961). Bays and funnel shaped estuaries have some of the highest tides in the world because the surge of rising water is confined into an increasingly narrow channel. The Severn estuary is a classic example of this. As it lies west – east, its mouth open to the vast Atlantic Ocean, when the moon swings across the sky, and the sun is pulling in the same line, the ocean swells in a vast following, shallow bulge which then washes up the ever-narrowing estuary. Many low lying coastal areas have vast areas of intertidal land whereas in other coastal areas, high cliffs emerge from deepish water and there is only a ceaseless rise and fall of tidal level against the cliff face.

In addition to coastal formation, tidal rhythms are affected by weather. Low atmospheric pressure might exaggerate the tidal rise, and high pressure suppresses it. Wind speed and direction can also either exaggerate or dampen tidal range. A *lee tide* occurs when tide flow and wind are in concert, and a *weather tide* when the wind blows in opposition to the tide. Storm tides, sometimes responsible for devastating floods (as in Eastern England in 1953), occur when high winds and low pressure combine with a high tide in such a way to pile water into coastal areas and estuaries.

In thinking about tidal rhythms it is important to note that times of high and low water do not synchronise with the 24-hour-day cycle in any simple way: '(B)ecause the earth rotates in relationship to the moon once every 24.8 hours, high tides occur on average every 12.4 hours' (Young, 1988: 27). Thus the timings

**Figure 14.1   Low tide at Sharpness Docks, Severn estuary (photograph by
the author) Note the figures for some idea of scale. At high tide
the water level will rise to the top of the dock wall**

of high and low water slowly migrate across the 24-hour grid. The other monthly
cycle of peaks (spring tides) and lows (neap tides) is 14.8 days.

As they rest on the Newtonian movements of sun, moon and earth, the *nominal*
times and heights of high and low water can be very accurately predicted into the
quite distant future (Carson, 1961). Standing on the sea walls built to defend the
coastal levels around the Severn estuary at the very highest tides I have always
marvelled at the confidence of the engineering as the water rises to within a few
inches of the top of the sea wall. A vast expanse of tidal, brackish water looms
some 4 metres over the protected land at the moment of high tide and would wash
far inland if unrestricted. Then the tide turns and slowly recedes to leave a draining
estuary. Six or so hours later (depending where you are) no water at all might
be visible, just a deep, indistinct perspective of sand, mud banks and draining
channels.

In the Severn estuary at the spring equinoctial tides, the difference between
high and low water is some 14 metres, the second highest tidal range in the world.
An impression of the difference in high and low water can be gained by looking
at dock infrastructure such as locks at low water (Figure 14.1). The force of this
natural process shapes the estuary and the life in and around it. A huge wedge of
rising water presses up the estuary, filling it up to the very brim of its sea defences

and washing back up the normally draining rivers, through both urban and rural landscapes. The famous Severn bore – a large tidal wave which washes up the lower reaches of the River Severn on days of the higher tides is a notable tourist attraction.

This dynamic, powerful, process presents huge challenges to processes of local government. As the Severn Estuary Partnership (SEP) put it:

> Britain's longest river brings vast quantities of water into the Severn estuary.
> Europe's biggest tide takes masses of water back up into the mainland. The
> mighty Severn influences the ways we live in many ways – and deserves all the
> attention we can give it! (SEP, 2005: 2)

Beyond the Severn estuary, the United Kingdom is an island nation with an estimated coast line of 17,820 km. All the intricacy and variety of that coast is washed by some of the highest tides in the world. Estuaries (which by definition are tidal) form a large part of the UK coastal geography and can contain vast areas of intertidal lands (including Morecambe Bay, 61506.22 ha. and The Wash, 107761.28 ha). The Nature Conservancy Council (1991) identified 155 estuaries around the British coast. They calculated that, 'the 9,320 km. of estuarine shoreline makes up 48 per cent of the longest estimate of the entire coast', and that '18,186,000 people live in large towns and cities adjacent to estuaries'. These landscapes, which can bring nature/wilderness into the heart of urban conurbations, or form our most remote wild landscapes, are distinguished by unique temporal and spatial characteristics which I explore below.

Tides and rhythms of life

I am interested in how tidal rhythms are folded into local processes, materialities and practices of places/landscapes – how tidal rhythms might distinctively pattern everyday life. Matless (2009) discusses studies of the wave patterns sculpted into the sands of estuarine intertidal landscapes, yet I argue that tides sculpt other material and lived patterns of life in coastal areas, creating differently different rhythmic traces in social and cultural geographies of work, recreation, community and individual identity.

This local temporal patterning of life contribute to what *Common Ground* (Clifford and King, 1993) call local distinctiveness, and what Thrift (1999) terms 'ecologies of place'. In these accounts however, it is the distinctive materialities, cultures and identities which are discussed. There is less attention to the distinctiveness of *temporal patterning*. Massey and Thrift (2003) and Massey (2005) have developed a more fully processual view which sees places and landscapes as outcomes of intersecting flows (material and non-material, human and non-human) coming together and coalescing into knowable forms. The temporal nature of these comings and goings and the rhythmic patterning

The *agencies* of tides and other *processes* and their associated rhythming and patterning, deserve our attention. Although attention is now being paid to the agencies of non-humans (Jones and Cloke, 2008) it is often as discrete entities (materials, organisms, machines) rather than processes.

Glimpses of tidal material culture

This is a watery globe with oceans covering some 70 per cent of its surface. Tides (of a sort) are ubiquitous phenomena, as all water responds to the gravity of the moon, though this only become obvious when it hits land. This means that they are ecologically highly significant. Watson (1973: 22) states that, 'every drop of water in the ocean responds to this force, and every living marine animal and plant is made aware of the rhythm'. Beyond these natural rhythms, all manner of material culture has been and is shaped by tidal rhythms. For example, boats which were designed to ply tidal coastal waters were often differently designed, having flat bottoms for settling on mud at low tide. Such distinctive boats as Severn Trows sailed coastal trade routes for centuries between small harbours set in tidal creeks and would have done so in rhythms and timings dictated by the estuary's fearsome tides.

In London, the great 19th century sewage engineering works, which saved the city from the worst excesses of pollution, were designed and (still) operate in conjunction with the tidal rhythms of the river Thames. Mains sewers were built parallel to the river to intercept existing sewage outfalls, and these ran east to a point where the sewage would be released when the tide of the river was falling, thus taking the outflow down-stream and away from the city. Thus the whole sewage network has a tidal rhythm. Similarly, the active nuclear power stations on the Severn estuary discharge low level radioactive polluted water into the estuary when the tide is ebbing, thus carrying it out into the Atlantic and away from the estuary's cities.

These are just a few of many possible examples of tidal material culture, which further includes intertidal spaces and their margins as key nature conservation sites; locations of heavy industry and civil infrastructure; places of work (ports, commercial fishing, coastal agriculture); and places of recreation (sailing, bird watching, recreational fishing, walking). Many of these practices have distinct temporal rhythms and distinctive material arrangements (sluices, swing bridges, piers, slipways, locks, jetties, steps leading to apparently nowhere, putcheon (tidal fishing traps). Being neither land nor sea, many exposed intertidal lands are peculiarly difficult to access. Often there are very specific local knowledges of how to access the spaces in terms of where, when and how (Robinson, 2007). They can be highly dangerous, and sites of labour exploitation, as the tragic case of the Chinese cockle pickers drowned by a rising tide in Morecambe Bay attests.

Rhythm and affective tidal spaces

Beyond the material cultures of tidal landscapes, I am interested in how the rhythms of tides get folded into affective practices/experiences of places/landscapes. On remote beaches, urban seaways and in vast estuarine landscapes, intertidal spaces and their land margins become highly potent, affective spaces of becoming which are generated and experienced in multiple ways. The impact of these landscapes has been remarked upon by Bill Adams,

> these places seem to have a very particular power. This lies in the sense of freedom that beaches offer, their sheer openness, and the novelty of the life they support ... They are places that literally have a life of their own, where rhythms of tides and seasons set an agenda that seems to stand outside human time. (1996: 2-3)

Tidal landscapes are affectively charged places because they stand at the edge of the sea, scramble that profound margin 'betwixt land and sea', create the liminal spaces of intertidal zones, threaten to inundate at high tide, threaten to drain to nothing at low tide, repeatedly empty and fill, and provide a temporality which is highly visible hour to hour. This then is a natural rhythm which is apprehended within *everyday* human time unlike other natural cycles such as that of the seasons, which usually only bring imperceptible change on a daily basis, and yet longer cycles, such as the growth and death of trees.

In what follows, I explore this idea drawing examples from literature and art. As I will show, numerous writers/artists use the turn of the tide and the relating transformation of space as affective settings for a whole range of narrative moments, experiences and feelings. Critical to the affective registers of tidal rhythms and their spaces is the fact that human life is a temporal process, one lived narratively over time which has progression from birth to death, but contains all manner of ebbs and flows and rhythms.

Tide, time and narrative

In literary narratives, high and low water are often used as symbols of beginnings and ends. The turn of the tide is used to locate 'us' and our stories in time – to mark a point where things can start, and things can end. This reflects a need not only for human stories to embed themselves in patterns of space and place, but also in rhythms of time. Just as landmarks and objects can be significant in terms of places (such as the prominent trees in Harrison, 1991) then *moments* and *periods* such as low tide, high tide, rising and falling tide, can become markers of the lived flow of time.

References to the tide occur in the opening and closing of such novels as Dickens's *Our Mutual Friend*, Eliot's *Mill on the Floss*, Du Maurier's *Frenchman's Creek*, Murdoch's *The Sea, The Sea*, Cary's *The Horse's Mouth*, Banville's *The Sea*,

Lynch's *The Highest Tide*, Burnside's *Glister* and Gallico's *The Snow Goose*. The flow and especially the turn of the tide, is a scene and mood setter, a symbol of some narrative about to begin, and then end. This motif is also used in travel/place writing, as in *The Kingdom by the Sea* by Paul Theroux and *Coasting* by Jonathan Raban.

A striking example is Joseph Conrad's seminal 20th century novel *Heart of Darkness*. The margins and time-space rhythms at the edge of land and sea are a key motif in Conrad's form of psychological realism. Two sets of his short stories are entitled *'Twixt Land and Sea* and *Within the Tides*. In *Heart of Darkness* Conrad sets the entire narration of the sombre tale on a boat anchored on the Thames estuary, between the turn of one tide and the next.

> (Opening) The Nellie, a cruising yawl, swung to her anchor without a flutter of the sails, and was at rest. The flood had made, the wind was nearly calm, and being bound down to the river, the only thing to do was to come to and wait for the turn of the tide. (1973: 5)

> (Close) Marlow ceased, and sat apart [...] in the pose of a meditating Buddha. Nobody moved for a time. "We have lost the first of the ebb", said the Director, suddenly. I raised my head. The offing was barred by a black bank of cloud, and the tranquil waterway leading to the uttermost ends of the earth flowed sombre under an overcast sky – seemed to lead into the heart of an immense darkness. (1973: 111)

Here, turn of the tide, the stillness, the vast looming space of the estuary and its sky, the proximity to the great city of Empire, London, act as a dramatic pause, even lacunae, in which the weight of the narrative and the horrors and uncertainties it portrays are framed to dramatic effect.

Key moments in human narratives are birth and death of self and others. A number of folklore sources tell how such moments in human and non-human lifecycles were believed to be affected by tidal rhythms. In *The Golden Bough* Fraser points out that,

> dwellers by the sea cannot fail to be impressed by the sight of its ceaseless ebb and flow, and are apt [...] to trace a subtle relation, a secret harmony, between its tides and the life of man, animals and plants. (1993: 34)

In the death scene of Mr Barkis in Dickens's *David Copperfield*, Mr Peggotty claims that:

> People can't die, along the coast ... except when the tide's pretty nigh out. They can't be born, unless it's pretty nigh in – not properly born, till flood. He's a going out with the tide. It's ebb at half-arter three, slack water half an hour. If he lives till it turns, he'll hold his own till past the flood, and go out with the next tide. (1996: 413)

Figure 14.2   **Beach art, Tenby: drawn upon the fresh sands after the**
previous tide and to be washed away by the next (Photograph
by the author)

Are such things just fancy? Watson (1973) tells of biological research showing that marine creatures such as shellfish can feel the pull of the moon and are physiologically in tune with it, even when removed from their normal environment. This link between the body and the pull of the moon is exemplified in a human sense by Walt Whitman, who when visiting American Civil War hospitals in Washington, felt that the wards, and the seriously wounded in them became calmer, (and died 'easier') in harmony with the turns of the tides (see Raban, 1992: 474).

Beyond the obvious (anti)climax of death, novelists use tides, particularly the Spring tides as psychogeographical moments of narrative climax. Two recent examples are John Banville's *The Sea*, and David Lynch's *The Highest Tide*. In *The Sea* the narrative as memoir takes an elliptical form where the dramatic, tragic denouement of the story, which takes place at an exceptionally high tide, is anticipated in the opening lines.

> They departed, the gods, on the day of the strange tide. All morning under a milky sky the waters in the bay swelled and swelled, rising to unheard-of heights. (Banville, 2005: 3)

In *The Highest Tide* the narrative opens with a discovery made by a boy while exploring intertidal land at a very low tide. The ebbing and flowing of the tide is then woven into the story which unfolds more optimistically as Lynch uses an exceptionally high tide to mark, and contribute to, the resolution of various strands of the story in the denouement. The use of tides in such works is unsurprising, precisely because it is such a monumental force and lives beside the sea are entangled with it!

There is a very obvious and literal way in which intertidal space feels new after the inundation of the previous high tide. On the sandy seaside beaches of Tenby and so many other resorts, the footprints, the sand graffiti (Figure 14.2), the sandcastles of one day are obliterated by the tide to leave the beach pristine again on the morrow. This cleaning of the marks of past occupation, and the knowledge that, for a while at least, this space was aquatic and deeply non-human, can make intertidal space feel fresh and new, a space of rejuvenation and euphoria, as the mark of the social is repeatedly washed away. This feeling is captured in the novel *Agnes Grey* by Anne Brontë:

> My footsteps were the first to press the firm, unbroken sands; nothing before me had trampled them since last night's flowing tide had obliterated the deepest marks of yesterday, and left it fair even, except where the subsiding water had left behind it the traces of dimpled pools and little running streams … Refreshed, delighted, invigorated, I walked along, forgetting all my cares, feeling as if I had wings on my feet… and experienced a sense of exhilaration to which I had been an entire stranger since the days of early youth. (1996: 145)

Conversely the draining and creeping tide and the unknowableness and openness of often vast intertidal landscapes can be a spatio-material medium for feelings of angst, threat, fear and dread. In George Crabbes' narrative tragic poem *Peter Grimes* the tidal marshes become a place of isolation and desolation.

> Thus by himself compelled to live each day,
> To wait for certain hours the tide's delay;
> At the same times the same dull views to see,
> [...]
> The water only, when the tides were high,
> When low, the mud-half covered and half-dry;
> [...]
> There anchoring, Peter chose from man to hide,
> There hang his head, and view the lazy tide
> In its hot slimy channel slowly glide (cited in Drabble, 1979: 73).

Similarly, Sylvia Plath found intertidal space raw and visceral – a place for sombre reflection:

Dawn tide stood dead low. I smelt
Mud stench, shell guts, gulls leavings.
Mussel Hunter at Rock Harbour (Plath, 1981)

To associate so many affective states with intertidal lands might imply that they are merely backdrops which are somehow socially constructed into different kinds of landscape. However I argue that tidal landscapes are stirring landscapes through ways which somehow penetrate the body/mind, not least through rhythm itself. In this way they might stir – or heighten emotional/affective states and thus might be fearful landscapes for some, euphoric landscape for others. The affective processes of tidal rhythms interpenetrate the rhythms of affective becoming in individual lives.

Performative practices of tidal rhythms

Sculptors, artists and photographers also explore intertidal spaces and their margins and examples abound of work that seeks to capture the rhythmic moods of tidal spaces in the representation of a moment – or series of moments. Turner painted estuaryscapes, Coster and MacDonald (1989) produced a wonderful photographic essay, *Man Made Wilderness,* about people using the 'beaches' of the Tees estuary at various states of the tide, alongside the heavy industrial sprawl of the area. Cornish artist Wilhelmina Barns-Graham created a series of delicate etchings of changing wave patterns of the sea – *Water Rhythm; Linear Movement (Incoming Tide Series).*

Here I am more interested in artists who work with tides in a more performative, material sense incorporate the space-time rhythms of the ebb and flow of tides into their artwork. Richard Long has produced a performative artwork which involved walking along the bank of the river Avon (UK), following the front of a rising tide, translating the rhythm and velocity of the tide into his own body. Simon Starling produced a hybrid performance/gallery work of a large piece of rock floated up the Avon on a wooden raft at high tide, this then being hoisted out of water in Bristol and placed in a gallery space. (http://www.spikeisland.org.uk/?q=exhibitions/long-starling). This echoes how the tide has been used for trade and transport, connecting Bristol to global sea trade routes and local coastal trading networks. It should not be forgotten that the slave ships which made Bristol a centre of the UK slave trade floated up the Avon on high tides, bearing their cargo of tightly packed, chained living souls. *Drift* by the Teri Rueb (http://www.terirueb.net/drift/index.html) is an artwork which employs locative ICT to allow and encourage people to 'wander among layered currents of sand, sea and interactive sounds that drift with the tides, and with the shifting of satellites as they rise and set, introducing another kind of drift'.

David Nash's work *Wooden Boulder* is an ecological/elemental art work that incorporates natural rhythmic processes over time. An oak boulder, three feet in diameter, was created in the late 1970s. Partly by accident it got stuck in a mountain stream in Wales. From then on, at times when the stream was in spate, the boulder was occasionally moved down stream. The artist began to chart this progress. Eventually the boulder found it way into the tidal reaches of the estuary of the River Dwyryd. According to Deakin (2007: 163) the boulder began to,

> wander the waters of the estuary, mysteriously disappearing up creeks, endlessly doubling back on itself in the ebb and flow, moving with each new tide, responding to the moon [...]. Nash went searching for it in a boat and lost it all together for a while [...]. During those chilly winter days of hide-and-seek he studied the tides and pored over charts, mapping the uncertain voyage. Then one January day the great oak apple reappeared on a saltmarsh and seemed almost settled for a moment until the equinox tide of 19 March 2003 floated it free. Nash watched from the boat [...] as the heavy sphere floated, most of its body submerged 'like a seal'. [...] It was just a far off dot when he last saw it on 30th of March. Somebody sighted *Wooden Boulder* floating close to the estuary's mouth a few days later, but it vanished in April 2003.

Perhaps the most significant 'tidal sculpture' in the UK is Anthony Gormley's *Another Place*. The piece consists of 100 life-size identical cast iron figures spaced along a two mile section of Crosby Beach in North West England. The figures all are orientated the same way, apparently looking out towards the sea horizon. They are placed at differing heights on the beach and at high tide are nearly all submerged, at low tide all exposed. In between, some figures are partially submerged, others at the water's edge and yet others standing clear of the water. The ceaseless, varying cycle of the tide becomes a critical component of the work. In a sense this work cannot be fully seen, as each tide, through day and night, and varying in heights from neap to spring, will bring a different extent and speed of inundation of the figures. The figures will also change colour each time they dry after submersion (according to a cyclical rhythm), and change over time (following a linear rhythm) as the sea, and sea creatures such as barnacles, affect their fabric.

Conclusion

I have shown in this chapter that the time-space rhythms of tides around the UK shore are bound into ecologies of place in many ways and that this produces affective relationships with place and landscape which inevitably dance to this fundamental, cyclical, but complex rhythm. How the rhythms of the tides interlink with other rhythmic processes of society has only been touched upon. Tides and their immediate effects can be a particularly powerful rhythm in coastal ecologies, partly because of their sheer scale and physical power.

However, tidal culture can come under threat. In Cardiff, the Taff estuary, with its large tides and mud banks, and wildlife designations, has been permanently flooded in order to create a cityscape of high real estate value. The old estuary

Cardiff Bay. Creating a superb waterfront environment.

Figure 14.3   The denial and obliteration of tidal culture. The Taff Barrage created permanent high water in the Taff estuary which was renamed Cardiff Bay (photograph by author)

has been represented as 'dormant' in contrast to 'vibrant' new dockland housing, but such an assessment serves only particular interests (Figure 14.3). Moreover, the NCC (1991) show how tidal areas of the Humber estuary (and many other estuaries) have been reduced through a series of land reclamations. The Severn estuary itself now faces the threat of a barrage for the purposes of tidal power generation – trapping high water up stream of the barrage and then releasing it through turbines. The tidal patterns and ranges will be severely affected, producing a landscape of necrosis (dead flesh), for in the tides is the life of this landscape. Indeed the prospect of sea level rise due to climate change poses a threat to many inter-tidal areas around the UK and beyond.

The moon, when first formed, was much closer to the Earth and tidal ranges and processes would have been much greater and more violent than they are now. This would have caused extreme levels of flooding, erosion and mixing of materials at the margins of land and sea. This possibly played a key part in generating suitable conditions for the creation of life on earth. We are in some respects hybrid in essence. The outcome of rhythmic mixings, and the rhythms of the tides may speak to us in very deep ways.

With thanks to the editor for his interest and guidance.

Chapter 15

Re-thinking Catastrophe in the Time of Climate Change

James Evans

The paradox

Nature is crazy. Nature is one big catastrophe. (Slavoj Žižek, 2008a)

We must reflect at the same time on the possibility of catastrophe and the responsibility, of possibly cosmic importance, that humanity bears to avert it. (Jean-Pierre Dupuy, 2005: 14)

Surely the most intriguing paradox of the early 21st century is that increasing certainty about impending climate catastrophe is not stimulating more action to prevent it. In this chapter I argue that this problem is one of perception, concerning our relationship to the planet on which we live and the future that we are bringing about.

Interestingly, the catastrophic dimension of nature has become well established in recent years, replacing the Romantic notion of nature as a balanced procession of cycles and rhythms. Revelations from the geological record show that mass extinctions have destroyed 99 per cent of the species that have ever lived (Gould, 2002). The plains of Northern America, upon which the pioneering ecologist Clements based his equilibrium theories of succession and climax, were actually the result of long-term burning by indigenous Indians (Deneven, 1992). Sudden catastrophes like hurricanes, tsunami, floods, forest fires and earthquakes, which disrupt nature's rhythms in the most extreme fashion, are accompanied by slower, less visible, but no less threatening catastrophes such as loss of polar ice cover and Colony Collapse Disorder amongst bees. Žižek's precedes the opening quote of this chapter with the irony that fossil fuels, which are driving the current climate catastrophe, were themselves the result of catastrophic losses of life in the past. His quote can thus be read literally and discursively; nature as 'environment' is crazy and catastrophic, but so too is nature as a cultural term. The disjuncture between the two has had disastrous consequences for our own efforts to solve environmental challenges.

The cultural baggage of the word nature has led some to advocate its wholesale abandonment. What we need, they say, is an 'ecology without nature'; a science purged of its antiquated romanticism (Morton, 2007). But the problem with this 'no-nature' position is that it invites a laissez-faire approach to the global environment. Nietzsche's primary concern was not to expose God as a myth, but with the dangers of doing so in the absence of a suitable replacement. So it is with nature today – while we have exposed the environmentalist myth of a harmoniously balanced nature upset by the arrival of industrialised humans and climate change, we have not yet found a suitable replacement. Accepting the catastrophic basis of existence upon this planet is necessary to cure ourselves of the romanticised notion of 'natural balance', but if by doing so we undermine the ethical imperative to act then we are in danger of throwing the baby out with the bathwater (Ereaut and Segnit, 2006). If there is no 'nature' to save, then what *are* we to do?

Part of the answer to this question, I argue, involves recovering the 'catastrophe' as part of the rhythm of existence, in order to ground a nature capable of reconnecting action with environmental concern. The following essay offers some initial thoughts in this vein, drawing upon the work of, amongst others, French philosopher Jean-Pierre Dupuy, to suggest that it is our conception of time itself which requires attention.

The post-politics of risk calculus

Catastrophes bring the protective function of the Hobbesian state into sharp focus. In his cultural history of catastrophes, Walter (2008) identifies the emergence of a knowledge society in Europe with first, the introduction of free information, and second, its ability to manage crises. As part of this progression, he argues, the Romantic era represented the period in which the first condition had been met, but effective crisis management had yet to emerge. In this period access to information generated massive technological and political advances, but society was unable to mitigate the negative excesses associated with them.

In the current 'knowledge society' the function of crisis management has been disbursed through the field of risk management, which casts hypothetical catastrophic events in the language of mathematical probability. Flood risk presents a classic example. Return periods measure the number of years in which a flood risk event of a specific magnitude might be expected to occur in a particular place. In light of these calculations, decision-makers might decide to construct flood defences to withstand a one in a hundred year flood event. The use of probabilities to manage risk is characteristic of all levels of environmental governance. While the Intergovernmental Panel for Climate Change has spent its last two reports attempting to distance itself from attaching exact probabilities to its statements (mostly because of the force that is subsequently attached to them by politicians (Shackley and Wynne, 1996)), it still deals in the nomenclature of 'likely', 'very likely' and so forth.

Dupuy argues that probabilities normalise risk as part of our accepted everyday world. For example, one of the few serious studies of nuclear risk estimated the chances of a Chernobyl-style nuclear reactor meltdown occurring in a German nuclear power plant with a forty year operational lifetime at 0.1 per cent. Given that there are over 150 nuclear power plants operating in the EU and over 400 globally, a simple extrapolation gives an estimate of a catastrophic accident occurring at 16 per cent in Europe and 40 per cent worldwide. There are assumptions here (for example, that German reactors are representative of most reactors), but it is not the uncertainty that underpins these calculations of risk that is the problem. It is irrelevant whether the chance of a reactor meltdown in Europe stands at 6 per cent, 16 per cent or 60 per cent in the next forty years – the probability alone can't tell us if this is too much. To make this judgement, society turns to the calculus of cost-benefit analysis (henceforth CBA).

CBA maximises overall utility by comparing the probable costs of disaster against the potential benefits that might accrue from adopting a certain technology. One argument against CBA is that counting individual preferences for various options fails to capture costs and benefits that accrue collectively over time. For example, most people would agree that the value of ancient woodland reflects more than simply the amount of money individual visitors would be willing to pay to visit it. Hence we can argue with Sagoff (1988) that CBA cannot fully capture higher values associated with the spiritual or historical meaning of life, because values are not reducible to simple economic preferences. Conversely, Žižek (2008b) argues that utilitarian calculations of self-interest like CBA *can* capture un-selfish motivations. The paradox of value-based (or altruistic) behaviour lies in the fact that it satisfies a preference of the individual to behave in a selfless way. The opposite of utilitarian self-interest is thus not altruism, but self-disinterest or 'resentment'. This leads Žižek (346) to suggest that those who argue against utilitarian decision-making tools like CBA on the grounds that they fail to capture 'higher' values are mistaken; 'what utilitarian ethics cannot properly account for is not the true good, but evil itself, which is ultimately against my long-term interests.' The environmental economist would retort that such valuation techniques are not intended to capture absolute values, but relative changes in costs and benefits. Indeed; but is there anything *relative* about the potentially catastrophic impacts of a nuclear meltdown or climate change?

A number of simple points support Žižek's position. First, it is practically impossible to capture the potential evils of future catastrophic scenarios economically. Nuclear power plants can only insure against approximately one per cent of the cost of a Chernobyl style meltdown, and are exempt from the legislation that requires full liability insurance for all other energy-producing installations. It is hard to place either a real or a projected cost on the horrific birth defects caused by Chernobyl. Costanza et al.'s (1997) efforts to value the Earth's ecosystem services lead to similar conclusions. Things like breathable air go from being plentiful enough to be effectively free to infinitely expensive as soon as they become unavailable. They are literally invaluable for life and thus very hard to 'cost' in any economic sense of the word. Second, and related to this, CBA applies a future discounting rate, which means that a cost or benefit now has more weight than a cost or benefit in the future. Discounting makes little sense in practical terms, as it may well be a cost to my future self that is incurred in the future. More worryingly, discounting future costs militates against taking the long-term view, which underpins the philosophy of sustainability to ensure that the rights of future generations are respected.

The precautionary principle is closely related to the logic of CBA, suggesting that humanity should progress cautiously in the face of risk, so as to obtain the benefits of adopting new technologies. The assumption underpinning the precautionary principle is that increases in knowledge will reduce the uncertainties associated with a particular course of action. This logic can also be used as a partial defence of the future discounting rate that is applied in CBA calculations, as increased knowledge is expected to mitigate future problems and produce as yet unforeseen benefits. But this logic assumes that the only form of uncertainty at play is epistemic, related to our knowledge – it takes no account of ontological uncertainty that is intrinsic to the event being considered. The IPCC (2007: 50) suggests 'key uncertainties are those that, if reduced, could lead to new robust findings,' and goes on to identify environmental feedback loops as the main unknown that prevents more 'robust' predictions. The possibility that some uncertainties are irreducible is not entertained. As Dupuy (2007: 2.1) states, the result of this is that a 'risk economist and an insurance theorist do not see and cannot see any essential difference between prevention and precaution and, indeed, reduce the former to the latter.' The failure to capture absolutely undesirable outcomes is based upon the conflation of epistemic and ontological uncertainty, with the result that 'in truth, one observes that applications of the 'precautionary principle' generally boil down to little more than a glorified version of 'cost-benefit' analysis' (ibid.).

Reduced to an economic calculation of risk, the catastrophe is removed from the arena of discussion, politically castrated and rendered as an object of safety and management (Keil and Glöckle, 2000). CBA and the precautionary principle generate a 'technical competence that does not 'think' – to use Heidegger's phrase – (and) that is the supreme danger' (Dupuy, 2007a: 243). Lack of political engagement around big questions like nuclear and climate change creates a democratic deficit, which disenfranchises people from the processes that impact upon their lives (Beck, 1992; Swyngedouw, 2008). In normalising the catastrophe these moves make it literally unthinkable. As Dupuy claims, 'contrary to what the promoters of the principle of precaution think, the cause of our non-action is not scientific uncertainty. We know it, but cannot make ourselves believe in what we know' (Dupuy, 2006: 147). Under these terms, environmental action becomes more an issue of credibility than a technical challenge. Applying this line of argument to global warming, Žižek (2008b: 454) suggests that 'with all the data regarding its nature, the problem is not the uncertainty about facts ... but our inability to believe that it can really happen: look through the window, the green grass and blue sky are still there, life carries on, nature follows its rhythm'. This inability to believe recalls the much discussed 'value-action gap', between what people say they do and what they actually do, which sticks like a thorn in the side of those whose job it is to mobilise the public to take environmental action. Risk management reduces the catastrophe to a probability attached to an unreal future, while the comforting rhythms of our own habitual existence makes it impossible to imagine one is really happening. Common sense and risk calculus conspire to depoliticise the catastrophe, and by extension action, replacing prevention with precaution. But why are we unable to believe what we know?

The accidental eschatology of historical time

The catastrophe conceived as probability remains caught within the frame of equilibrium. Take the notion of the 'tipping point' at which an environmental system will 'flip' (or, in the terminology of catastrophe theory, bifurcate) suddenly from one state into another, causing catastrophic system shifts. One of the best known examples involves the incremental warming of the North Atlantic, which may trigger massive cooling as melt-water from the Arctic floods across the surface of the Atlantic, shutting down the Gulf Stream. Although the latest IPCC (2007) report considers this unlikely in the 21st century, it remains perhaps the paradigmatic example of the tipping point in climate change science. The existence of tipping points is widely evidenced in the geological record, but tools such as CBA are entirely unable to deal with them. While relative changes in environmental systems are based upon extrapolations from predictable patterns, tipping points manifest inherent ontological uncertainty. In this sense, the tipping point is nature's equivalent to Žižek's notion of absolute evil that CBA is unable to capture, and which exposes the failings of utilitarianism. The return period upon which flood risk is based does not mean that a one in a hundred year flood will only occur once every century. The irony is that refusing to deal with the catastrophe on its own terms means that when it finally happens it can only be perceived as an *accident*; as the coming to pass of a future that was quite literally not meant to have been.

Paul Virilio (2007) argues that the accident is the defining feature of modern society, as any given technology implies its own accident. So the commercial airliner made the air disaster not only possible, but *inevitable*. It is this inevitability that leads Virilio to use the religious idea of 'eschatology' to describe humanity's headlong flight towards the end of the world. Risk management feeds off the logic of the accident, endeavouring to insulate humanity from the ultimate accident by simulating accidents. The Chernobyl disaster was caused by a safety training exercise to test the back-up systems designed to stabilise the core in the eventuality that the cooling rods failed. In this case, the simulation of the accident caused the accident (whereas the millions of computer simulations running of climate change can only cause an accident in the future through their predictions). Echoing J.G. Ballard's explorations of technology and violence (for example, 1973), Virilio's work is a meditation on why we, as a supposedly enlightened society, accept these technologies that both kill and have the potential to kill vast numbers of people so unquestioningly as part of our everyday lives. Virilio's (2007) eschatology concerns the increasing magnitude of the inevitable accident that accompanies the general speeding up of life, and finds its logical endpoint in the guise of climate change – the 'greatest accident of all time'.

The parallels between the nuclear accident that defined the twentieth century and the climate accident that is already defining the twenty-first are clear on many levels, not least the way in which the accident is already normalised through the *sacrifice* of part of the world's population who must bear the consequences of the inevitable catastrophe. To this extent, the calculations of risk enact a symbolic sacrifice of the few in the name of the progress of the many, raising some interesting parallels between utilitarianism and Girard's (1977) work on the function of sacrifice in modern society. In the age of climate change, the scapegoat, which pacifies those who cannot attain the material trappings of capitalism, takes the form of a symbolic sacrifice of so-many millions of Bangladeshis, who, scenarios tell us, may or may not be permanently submerged under the Indian Ocean in 50 years time. Jameson (2005) also discusses the psycho-religious function of apocalypse in modern culture, noting that the catastrophe operates metaphysically as a symbolic ending (or death) that holds the subsequent (but not always realised) promise of redemption or fulfilment.

Virilio sees our speeded up world as creating an inexorable march towards the end of time, and Dupuy argues that it is precisely this historical, linear, model of time that prevents humanity from fully perceiving the reality of future catastrophes. Historical time is familiar to us as a linear succession of moments, each of which is causally related to that which went before. Figure 15.1 shows the unfolding of historical time, with the future represented as a number of possible branches. The principle of the fixity of the past holds that what has happened can no longer be changed, and thus all that matters is the future. This familiar model of time's arrow is the temporal manifestation of causal rationality, whereby decisions are made at each fork on the basis only of future considerations. The classic example used by rational choice theorists concerns decisions that are made on the back of heavy losses. Say, for example, that I purchase shares that then halve in value. My subsequent decision to sell or not should not take any account of the losses incurred so far – this expense is a 'sunk cost' and belongs to the past – it is fixed now – and my decision to sell or not should be based purely on future projects.

The problem with historical time is that we can only react to the catastrophe after it occurs, because it is literally part of an 'unreal' future that hasn't happened yet. We can hypothetically re-draw Figure 15.1 in our heads, with a range of possible climate change predictions from the latest IPCC report attached to each of the four possible forks to show that each is made unreal by the very fact that it is only one possible route amongst many. Each is a prediction that does not have to come true. As Dupuy (2007; section 1) puts it, 'if the future is not real, it is not something that projects its shadow onto the present ... If the future is not real, there is nothing in it that we should fear, or hope for.' Because risk probabilities produce unreal futures, catastrophes assume the appearance of accidents to which we can only react after they have happened.

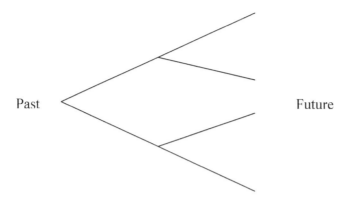

Figure 15.1   Historical time (after Dupuy, 2002: 184)

The error in this form of perception is highlighted by the simple fact that catastrophes *are* political. The Lisbon earthquake, which killed some 100,000 people, happened at the right time to prompt a widespread revolt against the prevailing theological beliefs of the time, while the Black Plague, which killed 20,000,000 in Europe some hundred years earlier, did not. Similarly, Hurricane Katrina and 9/11 fundamentally defined the 21st century political agenda of climate change and terrorism precisely because of *when* and *where* they occurred. The question is why we fail to perceive them as more than accidents.

Virilio (1997: 24-25) says of climate change that

> the time of the finite world is beginning ... general history has been hit by a new type of accident, the accident in its perception as visibly present – a cinematic and shortly digital perception that changes its direction, its customary rhythm, the rhythm of the ephemeredes or calendars – in other words the pace of the long time-span, promoting instead the ultra-short time-span.

This crisis in the perception of the future – 'the accident in its perception' – is precisely what Dupuy is taking aim at when he attacks the historical model of time upon which environmental decision-making tools like CBA are based. In order to prevent the catastrophe it is necessary to really believe that it can occur before it occurs. In Dupuy's (2005:.103) words, 'we must finish with the figure of the tsunami as a universal model of the disaster' and recover the catastrophe as a political subject. Despite all the bluster concerning the potential economic impacts of climate change created by reports such as the Stern Review (2006), this challenge is metaphysical, not economic. In order to believe and thus act, we need nothing less than 'a new notion of time' (Žižek, 2008:.459).

Projected time – a reason to act?

Returning to the example of an ill-advised share purchase, in reality, people tend to hold on to the shares hoping that they will recover, because selling the shares involves writing off the purchase cost and acknowledging what Dupuy (n.d.: 4) calls 'the *a posteriori* irrationality of the original decision' to purchase them. We behave in such a way so as to reassure ourselves that we didn't make an error in the past by imitating the consequences of different and more desirable choices that we might have made. This form of post-rationalisation appeals, because 'to act in a given way because what can be inferred from one's action is a favourable diagnosis of the state of the world, which is a state determined in the past, is to give oneself an unimaginable power: the power to *choose one's past*' (ibid: 6). But how can such a post-rationalisation be anything more than a deluded interpretation of one's past actions? And how can it enhance the utility of decision-making?

Dupuy takes the everyday example of a commodity exchange between two people, Peter and Mary, to indicate the value of thinking this way and the limitations of the causal rationalist model of historical time. In an ideal exchange, Peter would give Mary money, Mary would give Peter a commodity and both parties would benefit. If we apply purely rational logic to this exchange though, something quite strange happens. After Peter gives Mary the money, Mary would be compelled to keep it, without giving Peter the commodity in exchange. In historical time Mary only considers the future, and once she has the money, it makes sense for her to keep the money and the commodity. Further, as a rational actor, Peter would be able to predict, given that Mary is rational too, that she will behave in this way. As a result it would be rational for Peter not to instigate any kind of transaction and, in an outcome that echoes that of the Prisoner's dilemma, no exchange would ever be attempted and neither actor would benefit. In historical time the process of commodity exchange collapses under the weight of rational logic.

Dupuy's solution to this problem is to suggest that the way in which humans conceptualise time is in practice very different to the historical model suggested by causal rationality. He suggests that humans experience another form of time, which he calls the 'temporality of projects or projected time' (ibid: 13). Projected time lends itself to an entirely different form of rationality, and, as we shall see, ethics. Instead of holding the past as fixed and the future open, projected time fixes the future and opens the past. When the future is held constant it must be anticipated as the reaction in the present as an imagined past to an anticipated future event. Within this 'time of the project' the future and the past are linked in a loop (see Figure 15.2).

Considering the simple commodity exchange between Mary and Peter again, Dupuy shows how Mary can promise to cooperate in advance and follow through on this promise if she fixes the link between her intention to cooperate and subsequent cooperation. If the future is fixed (cooperation) then her promise in advance to cooperate is 'the inscription in the past of the future contingent: I will cooperate' (ibid: 25). This bootstrapping of future and past means that any world

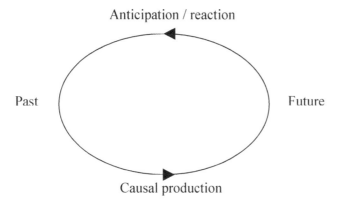

Figure 15.2   Project time (after Dupuy, 2002: 191)

in which Mary did not cooperate after Peter had given her the money would be a world in which she could not have previously promised to cooperate, and thus in which Peter would not have instigated the exchange. Within projected time, Mary knows that if she is given the chance to cooperate in the exchange then she must have promised to cooperate in all possible worlds. In this case, Mary has a good reason to genuinely intend to cooperate both in advance and during the transaction because 'if she does not do it, she forbids herself the benefits of the exchange by not giving Peter any reason to make the first move' (ibid.).

Two aspects of project time reward further consideration. First, this conceptualisation of time owes an obvious debt to Bergson, in that time is not conceived of as a linear succession of separate instances leaving a fixed past in their wake, but as a flow of moments which overlap with one another. In Bergson's famous musical metaphor, he suggests that perceiving moments in historical time is the equivalent of viewing music simply as a succession of individual notes, wherein each note follows separately from the last. By contrast, Bergson's (1960) conception of time seeks to capture the overall tune and rhythm that the notes make up. Removing or changing one note will alter the coherence of the overall tune, not just that part which is yet to come. In this way, events in the future can change the experience of that which went before as part of the rhythm of lived time. (Whether this is a matter of perception or an ontological quality of time remains potentially open, see Ansell-Pearson and Mularkey, 2002).

John O'Neill (2007) mounts a similar temporal critique of CBA, attacking the assumption of the additive separability of events and the associated irrelevance of the past on the grounds that they simply do not describe the way in which decisions are taken in the public realm. For him, non-commercial decisions do not take place from 'year zero' (87), but express a relation between the past and the future, that forms a 'social narrative'. For him, the utilitarianism of CBA smuggles a commercial temporality into social decision-making processes which simply does not produce decisions that are in the best human interest.

Dupuy's choice of subtitle for *Pour un catastrophisme éclairé*, 'when the impossible is certain', is telling, as he argues that we need 'an image of the future sufficiently catastrophic to be repulsive and sufficiently credible to trigger the actions that would block its realization' (2007b: 3.2). Dupuy (2002) also considers the problem whereby the successful prevention of a catastrophe like climate change would then invalidate the measures taken to avoid it. This problem of attribution is familiar to those who evaluate policies in the real world – how can we know what would have happened in their absence? Interestingly, the solution is to introduce an alternative notion of uncertainty into prevention, based not upon the forks of linear time, but on the simultaneous possibility that the catastrophe may still happen. Rather than reducing the impacts of catastrophe to a manageable calculus of scientific probability, all the while remaining within the rhythms of our own habitual existence, projected time allows the catastrophe to be perceived as *part of our time,* and thus recovered as a political subject whose desirability is open to debate, and whose occurrence we are free to prevent.

Returning to the commodity exchange example, Mary's reasoning requires her to step outside herself and infer the state of the world by observing her actions. Peter cooperates so she must have given him good reason to do so, and must thus cooperate. In this way she can act in such a way as to make her own past accord with a desirable vision of the future (i.e. that the mutually beneficial transaction takes place). By stepping outside of ourselves and observing the catastrophe as part of a real future, we have the power to act in such a way now as to stop it happening. This line of reasoning raises the second point of interest with respect to project time and current environmental problems; its ethical dimension. What is the exact nature of Mary's promise to cooperate? Mary's initial promise to cooperate holds no water in historical time, as the past should have no bearing over her subsequent decision to cooperate. Once Peter gives her the money she, as a rational being, will defect. In project time, however, Mary has committed herself to a transcendent rule of her own making, which applies in all possible worlds because it is held as a fixed, and thus inevitable, future.

Dupuy (n.d.: 26) calls this kind of promise to oneself a 'Kantian' promise, wherein an individual limits themselves by adhering to a transcendent rule in order to achieve some goal of self-improvement or enhancement. It should be noted that this is not the same as pre-committing oneself through some externally binding agreement, which would involve surrendering the freedom to choose. But further, within projected time, the Kantian promise loses its deontological appearance, and can become self-serving and even utilitarian. One example Dupuy uses is that of voting. In a democracy, one vote is unlikely to have any overall impact on the results of an election (especially in a system of non-proportional representation), and so a purely rational actor may decide that it is not worth the time and effort of actually researching each candidate's manifesto and going to vote. Rational behaviour would produce the 'free-rider effect' as individual voters depend on others making the effort to cast their vote. But the way in which voters actually think inverts this reasoning, corresponding far more accurately to project time. Individual voter thinks that 'if I vote, thus, I will be in a world in which those who would vote like me if they were to vote, will also vote. When I cast a ballot, I 'counterfactually' commit thousands, even tens of thousands, of similar votes' (ibid: 38). The ability to fix the future and use it as a basis for one's current actions is dependent upon the capacity to step outside of oneself and ask what would happen if one's own rule of action became a universal law. It is worth repeating that this position is logically incomprehensible within historical time – my actions in voting could have no causal effect on those of other people, making me in no way responsible for mobilising their vote. In project time, however, it becomes possible to attach a utility to my Kantian principle, asking 'what this hypothetical situation would bring it (me), in terms of its (my) self-interest ... if my maxim became a universal law?' (ibid: 37-38).

In the age of climate change, Kantian utilitarianism applies forcefully to a range of challenges that require us to act, ranging from recycling to binge flying. The citizens versus consumers debate misses the point that people engage in pro-environmental behaviour neither out of pure self-interest (in accordance with some utilitarian calculation of individual cost-benefit), nor out of pure 'duty' to some abstract principle or idea (for example, that it is 'right' to do so), but because they believe that such behaviour, if universalised, can bring about a more desirable (and less catastrophic) future.

The end of the end of the world

Paradoxically, then, it is only belief in the end of the world that can prevent the end of the world – only when we believe that the catastrophe is real can society act to prevent it. In the sense in which this chapter has articulated it, the challenge of reconstructing nature is a metaphysical one, reconceptualising nothing less than our notion of time. Dupuy (2002: 194) accepts that project time is a 'metaphysical fiction', but argues that historical time, *our time* in which we currently live and make decisions about the future, is no less fictional. For him, the job of metaphysics is to develop tools that help us to act. Project time disrupts the eschatology of the accident, allowing it to operate retroactively from the future as a cause for action rather than reaction, projecting the future into the past present. It becomes part of the unfolding rhythm of time, rather than an accidental disturbance or anomaly. In Žižek's (2008: 460) words, 'instead of saying "the future is open, we still have the time to act and prevent the worst", one should accept the catastrophe as inevitable, and then act to retroactively undo what is already written in the stars as our destiny'. In the age of climate change, project time may offer a more useful perceptual frame than the causal rationalist model of historical time. The loop is closed on the catastrophe, but in doing so it simultaneously provides an opening for hope, because our 'destiny has the status of an accident, of an error that we are free not to commit' (Dupuy, 2002: 216). The catastrophe does not stand in for a singular 'nature' that must be saved, or for a paralysing apocalypse that we are powerless to resist, but for a vision of a future society that *we can choose to prevent*.

Although he attributes it to 'someone else', Marxist literary critic Fredric Jameson (2005: 199) famously wrote that it is easier to imagine the end of the world than the end of capitalism. Dupuy's work sheds light on why this may be, by showing that the end of the world is itself articulated through utilitarian reasoning. IPCC predictions made within the open future of historical time show branches corresponding to what is generally perceived to be a catastrophic level of global warming of more than four degrees by the end of the twenty-first century. While we have critiqued this model of time for not making the catastrophe 'real' enough, it does at least make it possible. The risk of capitalism ending is not currently attached any probability, because the model of economic rationality embodied in historical time frames our entire perceptual apparatus. While it is possible to conceive of different environmental temporalities and rhythms (Evans and Jones, 2008; Jones and Evans, forthcoming), the logic of utilitarianism pervades even the 'sustainable' alternative – for example, what is recycling if not the attempt to squeeze yet more use out of a resource?

The incremental calculations of utility that take place in historical time have a limited ability to capture dramatic changes, whether they are associated with physical tipping points or absolute social evils. This is why historical time produces the catastrophe as accidental rather than systemic, and can only recover the catastrophe as a purely economic rather than a political object. Changing our relationship to the catastrophe is of major importance to those hoping to change the way in which society inhabits its environment. The moral barrenness of a society forged in the fire of utilitarianism has long been bemoaned (Hodgson, 1997); the prospect we are faced with in the environmental sphere is the more pertinent one of a barren planet.

Bibliography

Ackroyd, P. (2001), *London: The Biography* (London: Verso).

Adam, B. (1995), *Timewatch: The Social Analysis of Time* (Cambridge: Polity).

Adam, B. (1998), *Timescapes of Modernity* (London: Routledge).

Adams, B. (1996), *Future Nature* (London: Earthscan).

Adams, J. (1985), *Risk and Freedom: The Record of Road Safety Regulation* (Cardiff: Transport Publishing Projects).

Adams, J. (1995), *RISK* (London: Routledge).

Adams, J. (2004), 'Streets and the culture of risk aversion', in *What Are We Scared Of? The Value of Risk in Designing Public Space* (London: CABE).

Adey, P. (2008), 'Aeromobilities: geographies, subjects, vision', *Geography Compass* 2(5): 1318-36.

Adorno, T. (2001), *The Culture Industry* (London: Routledge).

Adorno, T. (2005), *Minima Moralia: Reflections on a Damaged Life*, trans. E.F.N. Jephcott (London: Verso).

AEBC (2007), *Mentality Principle* <http://www.aebc.com.au/articles/23/>, published 5 June 2007, accessed 30 January 2009).

Agnew, J. (2005), 'Space: place', in P. Cloke and R. Johnston (eds), *Spaces of Geographical Thought* (London: Sage).

Alheit, P. (1994), 'Everyday time and life time', in *Time and Society* 3(3): 305-319.

Allen, D. (2003), *Getting Things Done: The Art of Stress-Free Productivity* (London: Penguin).

Allen, D. (2006), 'I'm Halting My Personal Blog, For Now', available at <http://www.davidco.com/blogs/david/>, accessed 30 November 2008.

Allen, D. & Company (2008), 'Your Life is Your Investment: Leverage Your Investment', available at <http://www.davidco.com/individuals.php>, accessed 21 November 2008.

Allen, J. (1999), 'Worlds within Cities', in D. Massey et al. (eds), *City Worlds* (London: Open University).

Amin, A. and Thrift, N. (2002), *Cities: Reimagining the Urban* (Cambridge: Polity Press).

Anderson, B. (1983), *Imagined Communities* (London: Verso).

Anderson, B. and Tolia-Kelly, D. (2004), 'Matters in social and cultural geography', *Geoforum* 35: 669-674.

Anon. (2007), '20 Productive Ways to Use Your Free Time', available at <http://www.lifehack.org/articles/productivity/20-productive-ways-to-use-your-free-time.html>, accessed 21 November 2001.

Anon. (2007a), 'Simple Living Manifesto: 72 Ideas to Simplify Your Life', available at <http://zenhabits.net/2007/09/simple-living-manifesto-72-ideas-to-simplify-your-life>, accessed 5 October 2008.

Anon. (2008), 'Self Discipline', available at <http://howto.lifehack.org/wiki/Self_Discipline>, accessed 21 November 2008.

Ansell-Pearson, K. and Mullarkey, J. (2002), *Bergson: Key Writings* (London: Continuum).

Appadurai, A. (1990), 'Disjuncture and difference in the global cultural economy', *Theory, Culture and Society* 7: 295-310.

Aravot, I. (2002), 'Back to phenomenological placemaking', *Journal of Urban Design* 7(2): 201-212.

Arom, S. (1991), *African Polyphony and Polyrhythm: Musical Structure and Methodology* (Cambridge: Cambridge University Press).

Aslett, D. (2005), *The Office Clutter Cure* (Pocatello, Idaho: Marsh Creek).

Augé, M. (1995), *Non-Places: Introduction to an Anthropology of Supermodernity* (London: Verso).

Augoyard, J. and Torgue, H. (eds) (2005), *Sonic Experience: A Guide to Everyday Sounds* (London: McGill-Queen's University Press).

Axhausen, K., Zimmerman, A., Schonfelder, S., Rindsfuser, G. and Haupt, T. (2000), 'Observing the rhythms of daily life: a six-week travel diary', *Transportation* 29: 95-124.

Ballard, J.G. (2001), *Crash* (London: Picador).

Banville, J. (2005), *The Sea* (London: Picador).

Barker, C. (1999), *Television, Globalisation and Cultural Identities* (Buckingham: Open University Press).

Barthes, R. (1985), *The Responsibility of Forms: Critical Essays on Music, Art, and Representation* (New York: Hill and Wang).

Bartley, B. and R. Kitchin (eds), (2007), *Understanding Contemporary Ireland* (London: Pluto).

Basso, A. (2000), 'La Rumba ¿género de origen gangá?' *Clave* (1): 18-22.

Baudrillard, J. (1989), 'The anorexic ruins', in D. Kamper and C. Wulf (eds), *Looking Back on the End of the World* (New York: Semiotext(e)).

Bauman, Z. (2000), *Liquid Modernity* (Cambridge: Polity).

Bauman, Z. (1998), *Work, Consumerism and the New Poor* (Buckingham: Open University Press).

Beck, U. (1992), *Risk Society: Towards a New Modernity* (London: Sage).

Bendixson, T. (1974), *Instead of Cars* (London: Temple Smith).

Benjamin, W. (2002), *The Arcades Project*, trans. by H. Eiland and K. McLaughlin (Cambridge, MA: Harvard University Press).

Bergson, H. (1960), *Time and Free Will* (New York: Harper & Row).

Bergson, H. (1999), *Duration and Simultaneity: Bergson and the Einsteinian Universe* (Manchester: Clinamen Press).

Billig, M. (1995), *Banal Nationalism* (London: Sage).

Bissell, D. (2007), 'Animating suspension: waiting for mobilities', *Mobilities* 2(2): 277-298.

Blanco, J. (2000), 'Rumba, la fiesta Cubana', *Salsa, Revista Cubana* 13-16.

Boenisch, P. (2007), 'Aesthetic art to aisthetic act: theatre, media, intermedial performance', in F. Chapple and C. Kattenbelt (eds), *Intermediality in Theatre and Performance* (Amsterdam, New York: Rodopi).

Bonham, J. (2006), 'Transport: disciplining the body that travels', *Sociological Review* 54(1): 57-74.

Borden, I. (2001), *Skateboarding, Space and the City: Architecture and the Body*, (Oxford: Berg).

Bourdieu, P. (1986), *Distinction: A Social Critique of the Judgement of Taste* (London: Routledge).

Bourdin, A. (2003), Workshop 4: 'hypermodern individuals?', International Colloquium, *The Sense of Movement: Modernity and Mobilities in Contemporary Urban Societies*, organised by IVM, Centre culturel international de Cerisy-la-Salle, Paris, France, Institute pour la ville en mouvement.

Bowman, W. (2006), 'Musical experience as aesthetic: What cost the label?', *Contemporary Aesthetics Journal* Vol. 4, available at <http://www. contempaesthetics.org/newvolume/pages/journal.php?volume=16>, accessed 2 March 2009.

Boyle, P. (2004), 'Mapping the Lines: An Exploration of Mobility and Urban Spaces Amongst Bicycle Couriers' (unpublished MA thesis, University of Windsor, Ontario).

Bristow, G. and Morgan, K. (2006), 'De-industrialization, the new service economy and the search for post-industrial prosperity', in A. Hooper and J. Punter (eds), *Capital Cardiff 1975-2020: Regeneration, Competitiveness and the Urban Environment* (Cardiff: University of Wales Press).

British Eventing (2009), *Rules and Members Handbook* (Stonleigh: British Eventing).

Brontë, A. [1847] (1996), *Agnes Grey* (London: Wordsworth Classics).

Brooks-Pazmany, K. (1983), *United States Women in Aviation, 1919-1929* (Washington, DC: Smithsonian Institution).

Brose, H-G. (2004), 'An introduction towards a culture of non-simultaneity', in *Time and Society* 13(1): 5-26.

Bull, M. (2000), *Sounding out the City* (Oxford: Berg).

Burgess-Wise, D. (2001), *Ford at Dagenham: The Rise and Fall of Detroit in Europe* (Derby: Breedon Books).

Buttimer, A. (1976), 'Grasping the dynamism of lifeworld', *Annals of the Association of American Geographers* 66(2): 277-292.

Byrne, D. (2001), *Understanding the Urban* (Basingstoke: Palgrave).

Callon, M. and Law, J. (1995), 'Agency and the Hybrid Collectif', *The South Atlantic Quarterly* 94, 2, 481-507.

Capatti, A. (1996), 'In praise of rest', *Slow* 24 (October): 5-7.

Carlson, M. (1996), *Performance: A Critical Introduction* (London: Routledge).

Carlstein, T., Parkes, D. and Thrift, N. (eds) (1978a), *Making Sense of Time* (London: Edward Arnold).

Carlstein, T., Parkes, D. and Thrift, N. (eds) (1978b), *Human Activity and Time Geography* (London: Edward Arnold).

Carson, R.L. (1952), *Under the Sea-Wind* (London: Staples Press Limited).

Carson, R.L. (1961), *The Sea Around Us* (Oxford: Oxford University Press).

Cartwright, D.E. (2000), *Tides: A Scientific History* (Cambridge: Cambridge University Press).

Castree, N. (2009), 'The spatio-temporality of capitalism', *Time and Society* 18(1): 26-61.

Central Statistics Office (CSO) (2003), *Measuring Ireland's Progress – Indicators Report Volume 1* (Dublin: Stationery Office).

Central Statistics Office (CSO) (2007), *Measuring Ireland's Progress* (Dublin: Stationery Office).

Clifford, S. and King, A. (eds), (1993), *Local Distinctiveness: Place, Particularity and Identity* (London: Common Ground).

Cloke, P., Philo, C. and Sadler, D. (1991), *Approaching Human Geography* (London: Paul Chapman).

Cirillo, F. (2007), 'The Pomodoro Technique (The Pomodoro)', available at <http://www.pomodorotechnique.com/resources/cirillo/ThePomodoroTechnique_v1-3.pdf>, accessed 7 December 2008.

Cirillo, F. (no date), 'The Pomodoro Technique: Eliminate the Anxiety of Time; Enhance Focus and Concentration', available at: <http://www.pomodorotechnique.com>, accessed 7 December 2008.

Clarke, E. (2006), 'Shared space – the alternative approach to calming traffic', TEC, September, available at: <www.hamilton-baillie.co.uk/papers/14_tec.pdf>, accessed January 2007.

Colebrook, C. (2002), 'The politics and potential of everyday life', *New Literary History* 33: 687-706.

Conlon, D. (2007), 'The nation as embodied practice: Women, migration and the social production of nationhood in Ireland', PhD thesis, Department of Environmental Psychology, Graduate Center, City University of New York.

Conrad, J. (1973), *Heart of Darkness* (Harmondsworth: Penguin Books).

Corcoran, M. (2006), 'Ethno-city', in M. Corcoran and M. Peillon (eds), *Uncertain Ireland: A Sociological Chronicle 2003-2004* (Dublin: Institute of Public Administration), 181-194.

Costanza, R. et al. (1997), 'The value of the world's ecosystem services and natural capital', *Nature* 38(7): 253-260.

Coster, G. and MacDonald, I. (1989), 'Man-Made Wilderness', *Independent Magazine*, 23-27.

Covey, S.R. (1990), *The 7 Habits of Highly Effective People* (New York: Free Press).

Crang, M. (2000), 'Urban morphology and the shaping of transmissable city', *City* 4(3): 303-315.

Crang, M. (2001), 'Rhythms of the city: temporalised space and motion', in J. May and N. Thrift (eds), *Timespace: Geographies of Temporality* (London: Routledge).

Cresswell, T. (2004), *Place: A Short Introduction* (Oxford: Blackwell Publishing).

Cresswell, T. (2006), *On the Move: Mobility in the Modern Western World* (London: Routledge).

Cresswell, T. (2006), '"You cannot shake that shimmie here": Producing mobility on the dance floor', *Cultural Geographies* 13: 55-77.

Cronon, W. (1995), 'The trouble with wilderness; or, getting back to the wrong nature', in W. Cronon (ed.), *Uncommon Ground: Rethinking the Human Place in Nature* (New York: W.W. Norton & Co.).

Cronin, A. (2006), 'Advertising and the metabolism of the city: urban space, commodity rhythm', *Environment and Planning D: Society and Space* 24: 615-632.

Crouch, D. (2000), 'Places around us: embodied lay geographies in leisure and tourism', *Leisure Studies* 19: 63-76.

Csikszentmihalyi, M. and Csikszentmihalyi, I.S. (eds) (1988), *Optimal Experience: Psychological Studies of Flow in Consciousness* (Cambridge: Cambridge University Press).

Csikszentmihalyi, M. (2004), *Good Business: Leadership, Flow, and the Making of Meaning* (London: Penguin).

Cwerner, S. (2006), 'Vertical flight and urban mobilities: the promise of reality of helicopter travel', *Mobilities* 1(2): 191-215.

Daniel, Y. (1995), *Rumba: Dance and Social Change in Contemporary Cuba* (Bloomington, IN: Indiana University Press).

Deakin, R. (undated), *Wildwood* (London: Hamish Hamilton).

De Certeau, M. (2002), *The Practice of Everyday Life* (Berkeley: University of California Press).

Degen, M., Rose, G. and Basdas, B. (2009), 'Bodies and everyday practices in designed urban environments', *Science Studies*, under review.

Degen, M. (2008), *Sensing Cities: Regenerating Public Life in Barcelona and Manchester* (Routledge: London).

Delaney, E. (2004), 'The vanishing Irish? The exodus from Ireland in the 1950s', in D. Keogh, F. O'Shea and C. Quinlan (eds), *Ireland: The Lost Decade in the 1950s* (Cork: Mercier Press), 80-86.

Deleuze, G. and Guattari, F. (1987), *A Thousand Plateaus: Capitalism and Schizophrenia* (Minneapolis: University of Minnesota Press).

Deneven, W. (1992), 'The pristine myth: the landscape of the Americas in 1492', *Annals of the Association of American Geographers* 82: 369-385.

Derrida, J. (1976), *Of Grammatology*, trans. Gayatri Spivak (Baltimore, MD: Johns Hopkins University Press).

Desmond, J. (1997), 'Embodying difference: issues in dance and cultural studies' and 'introduction', in J. Desmond (ed.), *Meaning in Motion* (London: Duke University Press).

Dewey, J. (1934), *Art as Experience* (London: George Allen & Unwin).

Dewsbury, J.D., Harrison, P., Rose, M. and Wylie, J. (2002), 'Introduction: enacting geographies', *Geoforum* 33: 437-440.

Dickens, C. (1996), *David Copperfield* (London, Penguin Classics).

Dodgson, R. (2008), 'Geography's place in time', *Geografiska Annaler* 90(1): 1-15.

Drabble, M. (1979), *A Writer's Britain: Landscape in Literature* (London: Thames and Hudson).

Dufrenne, M. (1973), *The Phenomenology of Aesthetic Experience* (Evanston, IL: Northwestern University Press).

Duneier, M. (1999), *Sidewalk* (New York: Farrar, Straus & Giroux).

Dupuy, J-P. (n.d.), *Time and Rationality* available at <http://www.cgm.org/Forums/Confiance/index.html>, accessed 5 February 2009.

Dupuy, J-P. (2002), *Pour un catastrophisme éclairé: Quand l'impossible est certain* (Paris: Seuil).

Dupuy, J-P. (2005), *Petite métaphysique des tsunamis* (Paris: Seuil).

Dupuy, J-P. (2006), *Retour de Chernobyl* (Paris: Seuil).

Dupuy, J-P. (2007a), 'The catastrophe of Chernobyl twenty years later', *Estudos Avançados* 21(59): 243-52.

Dupuy, J-P. (2007b), 'Rational choice before the apocalypse', *Anthropoetics – The Journal of Generative Anthropology* 13(3), available at: <http://www.anthropoetics.ucla.edu/ap1303/>, accessed 12 February 2009.

Edensor, T. (2000), 'Moving through the city', in D. Bell and A. Haddour (eds), *City Visions* (London: Palgrave).

Edensor, T. (2005), *Industrial Ruins: Space, Aesthetics and Materiality* (Oxford: Berg).

Edensor, T. (2006), 'Reconsidering national temporalities: institutional times, everyday routines, serial spaces and synchronicities', *European Journal of Social Theory* 9(4): 525-545.

Edensor, T. and Holloway, J. (2008), 'Rhythmanalysing the coach tour: The ring of Kerry, Ireland', *Transactions of the Institute of British Geographers* 33: 483-501.

Edensor, T. (2009), 'Mobility, rhythm and commuting,' in T. Cresswell and P. Merriman (eds), *Mobilities: Practices, Spaces, Subjects* (Aldershot: Ashgate).

Elchardus, M. and Smits, W. (2006), 'The persistence of the standardised life cycle', in *Time and Society* 15(2/3): 303-326.

Elden, S. (2004a), 'Rhythmanalysis: An introduction', in S. Elden and G. Moore (eds), *Henri Lefebvre: Rhythmanalysis: Space, Time and Everyday Life* (Aldershot: Ashgate).

Elden, S. (2004b), *Understanding Henri Lefebvre: Theory and the Possible* (London: Continuum).

Ellegård, K. and Vilhelmson, B. (2004), 'Home as a pocket of local order: everyday activities and the friction of distance', in *Geografiska Annaler* 86B(4): 281-296.

Ereaut, G. and Segnit, N. (2006), 'Warm words: how are we telling the climate story and can we tell it better?', available at: <http://www.ippr.org.uk/members/download.asp?f=%2Fecomm%2Ffiles%2Fwarm%5Fwords%2Epd>, accessed 11 November 2007.

Evans, G. (2003), 'Hard branding the cultural city – From Prado to Prada', *International Journal of Urban and Regional Research* 27(2): 417-440.

Evans, J. and Jones, P. (2008), 'Towards Lefebvrian socio-nature? A film about rhythm, nature and science', *Geography Compass* 2(3): 659-670, also available at:<http://www.blackwell-compass.com/subject/geography/article_view?article_id=geco_articles_bpl107>, accessed 20 February 2009.

Evening News, The (12 March 1938), '"Compulsory crossings" test mirrors London of the future', 5.

Evening Standard, The (27 October 1931), 'When walkers control traffic', 10.

Examination of Nuclear Meltdowns. Study commissioned by the German Federal Ministry for Science and Technology (Rheinland: TÜV).

Featherstone, M., Thrift, N. and Urry, J. (eds) (2005), *Automobilities* (Thousand Oaks, CA: Sage).

Felski, R. (2002), 'Introduction', *New Literary History* 33: 607-22.

Flaherty, M. (1998), *A Watched Pot: How We Experience Time* (New York: New York University Press).

Fincham, B. (2006), 'Bicycle messengers and the road to freedom', *Sociological Review* 54(1): 208-222.

Fischer, F. (2003), 'Risk assessment and environmental crisis: toward an integration of science and participation' in S. Campbell and S. Fainsten (eds), *Readings in Planning Theory* (Oxford: Blackwell).

Flusty, S. (2000), 'Thrashing Downtown: play as resistance to the spatial and representational regulation of Los Angeles', *Cities* 17(2): 149-158.

Fiore, N. (2007), *The Now Habit: A Strategic Program for Overcoming Procrastination and Enjoying Guilt-free Play* (Los Angeles: Tarcher).

Fiore, N. (2008), 'Now Habit Schedules', available at: <http://www.neilfiore.com/nowhabit.shtml>, accessed 5 November 2008.

43 Folders (2007), 'Moleskine Friendly Fountain Pens', available at: <http://wiki.43folders.com/index.php/Moleskine_Friendly_Fountain_Pens>,accessed 20 April 2008.

Foster, R. and Kreitzman, L. (2004), *Rhythms of Life: The Biological Clocks that Control the Daily Lives of Everything Living Thing* (London: Profile Books).

Foster, S. (1997), 'Dancing bodies', in J. Desmond (ed.), *Meaning in Motion* (London: Duke University Press).

Foucault, M. (1991/1975), *Discipline and Punish: The Birth of the Prison*, trans. A. Sheridan (Harmondsworth: Penguin).

Franklin, A. and Evans, R. (2009), 'From work horse to hobby horse: exploring the micro geographies of equestrian pursuits within the British countryside', *Journal of Rural Studies* (forthcoming).

Fraser, J. (1993), *The Golden Bough* (Ware: Wordsworth Reference).

Frykman, J. and Löfgren, O. (eds) (1996), 'Introduction', *Forces of Habit: Exploring Everyday Culture* (Lund: Lund University Press).

Fuente, A. (2001), *The Resurgence of Racism in Cuba*, NACLA Report on the Americas, May/June: 29-34.

Fundació Tot Raval (2005), *Memoria 2005* (Barcelona: Fundació Tot Raval).

Galison, P. (1994), 'The ontology of the enemy: Norbert Wiener and the cybernetic vision', *Critical Inquiry* 21(1): 228-266.

Gardiner, M. (2000), *Critiques of Everyday Life* (London: Routledge).

Gardiner, M. (2004), 'Everyday utopianism – Lefebvre and his critics', *Cultural Studies* 18(2/3): 228-254.

Gehl, J. (2001), *Life Between Buildings* (Arkitektens Forlag: Danish Architectural Press).

Gieryn, T. (2000), 'A space for place in sociology', *Annual Review of Sociology* 26: 463-496.

Girard, R. (2005), *Violence and the Sacred*, trans. P. Gregory (London: Continuum).

Gottdeiner, M. (2001), *Life in the Air: Surviving the New Culture of Air Travel* (Boston: Rowman and Littlefield).

Gould, S. (2002), *The Structure of Evolutionary Theory* (Cambridge, MA: Harvard University Press).

Graham, S. and Thrift, N. (2007), 'Out of order: Understanding repair and maintenance', in *Theory, Culture & Society* 24(3): 1-25.

Gray, B. (2003), *Breaking the silence: Emigration, gender and the making of Irish cultural memory*, Working Paper Series, WP 2003-02, Limerick: Department of Sociology, University of Limerick.

Gray, B. (2004), *Women and the Irish Diaspora* (London: Routledge).

Gray, B. (in prep.), 'The non-migrant subject: awaiting a viable modern subjectivity', Limerick: Department of Sociology, University of Limerick.

Gren, M. (2001), 'Time-geography matters', in J. May and N. Thrift (eds), *Timespace: Geographies of Temporality* (London: Routledge).

Grosz, E. (1994), *Volatile Bodies: Theories of Representation and Difference* (Bloomington, IN: Indiana University Press).

Grosz, E. (1998), 'Bodies-cities', in H. Nast and S. Pile (eds), *Places Through the Body* (London: Routledge).

GRS (1990), *Deutsche Risikostudie Kernkraftwerke. Fachband 5. Untersuchung von Kernschmelzunfällen* [German Risk Study of Nuclear Power Plants – Vol. 5].

GTDtv™ (2006), *David Allen Promo*, available at: <www.davidco.com/video/index.php>, accessed 20 November 2008.

Hägerstrand, T. (1970), 'What about people in regional science?', *Papers in Regional Science* 24(1): 7-21.

Hagerstrand, T. (1977), 'The impact of social organization and environment upon the time-use of individuals and households' in A. Kuklinski (ed.), *Social Issues and Regional Policy and Regional Planning* (The Hague and Paris: Mouton).

Hagman, O. (2006), 'Morning queues and parking problems: on the broken promises of the automobile', *Mobilities* 1(1): 63-74.

Hall, S. (1996), 'Gramsci's relevance for the study of race and ethnicity', in D. Morley and K. Hsing Chen (eds), *Stuart Hall; Critical Dialogues in Cultural Studies* (London: Routledge).

Hallam, E. and Ingold, T. (2007), 'Creativity and Cultural Improvisation: An Introduction', in E. Hallam and T. Ingold (eds), *Creativity and Cultural Improvisation* (London: Routledge).

Hancock, B. (2005), 'Steppin' out of whiteness', *Ethnography* 6(4): 427-461.

Hannam, K., Sheller, M. and Urry, J. (2006), 'Mobilities, immobilities, and moorings', *Mobilities* 1(1): 1-22.

Hannigan, J. (2004), 'Boom towns and cool cities: the perils and prospects of developing a distinctive urban brand in a global economy', paper presented at the Leverhulme International Symposium: The Resurgent City, London School of Economics.

Haraway, D. (2003). *The Companion Species Manifesto: Dogs, People and Significant Otherness* (Chicago: Prickly Paradigm Press).

Haraway, D. (2004), 'A manifesto for cyborgs: science, technology, and socialist feminism in the 1980s', in D. Haraway (ed.), *The Haraway Reader* (London: Routledge).

Harootunian, H. (2004), 'Shadowing history – national narratives and the persistence of the everyday', *Cultural Studies* 18(2/3): 181-200.

Harrington, J. (2005), 'Citizenship and the biopolitics of post-nationalist Ireland', *Journal of Law and Society* 32(3): 424-449.

Harrison, F. (1991), *The Living Landscape* (London: Mandarin Paperbacks).

Harrison, P. (2000), 'Making sense: embodiment and the sensibilities of the everyday', *Environment and Planning D: Society and Space* 18: 497-517.

Harvey, D. (1989), *The Condition of Postmodernity* (Oxford: Blackwell).

Harvey, D. (1996), *Justice, Nature, and the Geography of Difference* (Oxford: Blackwell).

Hensley, S. (2008), *The Embodiment of Rumba in Cuba* (Milton Keynes: The Open University (unpublished thesis).

Herzfeld, M. (1997), *Cultural Intimacy: Social Poetics in the Nation-State* (London: Routledge).

Highmore, B. (2002), 'Street life in London: towards a rhythmanalysis of London in the late nineteenth century', in *New Formations* 24: 171-193.

Highmore, B. (2004), 'Routine, social aesthetics and the ambiguity of everyday life', *Cultural Studies* 18 (2/3): 306-327.

Hill, J. (2003), *Actions of Architecture: Architects and Creative Users* (London: Routledge).

Hodgson, G. (1997), 'Economics, environmental policy and the transcendence of utilitarianism', in J. Foster (ed.), *Valuing Nature?* (London: Routledge).

Holliday, R. and Jayne, M. (2000), 'The Potters' holiday', in Edensor, T. (ed.), *Reclaiming Stoke-on-Trent: Leisure, Space, and Identity in The Potteries* (Stoke-on-Trent: Staffordshire University Press).

hooks, b. (1997), 'Selling hot pussy: representations of black female sexuality in the cultural marketplace', in K. Conboy, N. Medina and S. Stanbury (eds), *Writing on the Body* (New York: Columbia University Press).

Horton, D. (2006), 'Environmentalism and the bicycle', *Environmental Politics* 15(1): 41-58.

Howes, D. (2005), 'Hyperesthesia, or the Sensual Logic of Late Capitalism', in D. Howes (ed.), *The Empire of the Senses* (Oxford: Berg).

Imrie, R. and Thomas, H. (1997), 'Law, legal struggles and urban regeneration: rethinking the relationships', *Urban Studies* 34(9): 1401-1418.

Ingold, T. (1993), 'The temporality of the landscape', *World Archaeology* 25(2): 24-174.

Ingold, T. (2000), *The Perception of the Environment: Essays in Livelihood, Dwelling and Skill* (London: Routledge).

Intergovernmental Panel on Climate Change (2007), *Climate Change 2007: Synthesis Report* available at: <http://www.ipcc.ch/ipccreports/ar4-syr.htm>, accessed 20 February 2009.

Jackson, P. (2004), *Inside Clubbing: Sensual Experiments in the Art of Being Human* (Oxford: Berg).

Jacobs, J. (1961), *The Death and Life of American Cities* (Harmondsworth: Pelican).

Jacobs, J. and Nash, C. (2003), 'Too little, too much: cultural feminist geographies', *Gender, Place and Culture* 10(3): 265-279.

Jameson, F. (1991), *Postmodernism or the Cultural Logic of Late Capitalism* (London: Polity).

Jameson, F. (2003), 'The end of temporality', *Critical Enquiry* 29: 695-718.

Jameson, F. (2005), *Archaeologies of the Future: The Desire Called Utopia and Other Science Fictions* (London: Verso).

Jain, J. (2006), 'By passing and WAPing: reconfiguring timetables for "real-time" mobility in mobile technologies', in M. Sheller and J. Urry (eds), *Mobile Technologies of the City* (London: Routledge).

Jarvis, H. (2005), 'Moving to London time: household co-ordination and the infrastructure of everyday life', in *Time and Society* 14(1): 133-154.

Jenkins, R. (2002), 'In the present tense: time, identification and human nature', *Anthropological Theory* 2(3): 267-280.

Jiron, P. (2007), 'Place making in the context of urban daily mobility practices: actualising time space mapping as a useful methodological tool', in E. Huijbens (ed.), *Sensi/able Spaces – Space, Art and the Environment* (Cambridge: Cambridge Scholars Press).

Jiron, P. (2008), 'Mobility on the move: examining urban daily mobility practices in Santiago de Chile', London School of Economics and Political Science, PhD thesis in Urban and Regional Planning.

Johnston, D. (2006), 'DIY Planner Hipster PDA Edition', available at: <http://www.diyplanner.com/templates/official/hpda>, accessed 20 April 2008.

Jonas, H. (1984), *The Imperative of Responsibility: In Search of Ethics for the Technological Age* (Chicago: University of Chicago Press).

Jones, O. (2005), 'An emotional ecology of memory, self and landscape', in J. Davidson, L. Bondi and M. Smith (eds), *Emotional Geographies* (Aldershot: Ashgate).

Jones, O. and Cloke, P. (2008), 'Non-human agencies: trees in place and time', in C. Knappett and Malafouris, L. (eds), *Material Agency: Towards a Non-Anthropocentric Approach* (New York: Springer).

Jones, P. and Evans, J. (2006), 'Time for sustainability: exploring time, the city and non-humans', University of Birmingham Working Paper 67.

Jones, R. (2003), 'Multinational investment and return migration in Ireland in the 1990s – a country-level analysis', *Irish Geography* 36(2): 153-169.

Julier, G. (2000), *The Culture of Design* (London: Sage).

Kaplan, C. (2006), 'Mobility and war: the cosmic view of US air power', *Environment and Planning A* 38(2): 395-407.

Kärrholm, M. (2009), 'To the rhythm of shopping: on synchronisation in urban landscapes of consumption', in *Social and Cultural Geography* 10(4): 421-440.

Kaufmann, V. (2002), *Re-thinking Mobility and Contemporary Sociology* (Aldershot: Ashgate).

Keil, D. and Glöckle, W. (2000), 'Sicherheitskultur im Wettbewerb – Erwartungen der Aufsichtsbehörde', *Atomwirtschaft* 45(10): 588-592.

Klein, N. (2008), *The Shock Doctrine: the Rise of Disaster Capitalism* (London: Picador).

Kofman, E. and Lebas, E. (eds) (1995), 'Introduction', in *Henri Lefebvre: Writings on Cities* (Oxford: Blackwell).

Korsmeyer, C. (2004), *Gender and Aesthetics* (London: Routledge).

Krause, B. (1993), 'The Niche hypothesis: a hidden symphony of animal sounds, the origins of musical expression and the health of habitats', *The Explorers Journal* Winter: 156-160.

Kroker, K. (2007), *The Sleep of Others and the Transformations of Sleep Research* (Toronto: University of Toronto Press).

Kutzinski, V. (1993), *Sugar's Secrets: Race and the Erotics of Cuban Nationalism* (Charlottesville: University Press of Virginia).

Labelle, B. (2008), 'Pump up the bass: rhythm, cars and auditory scaffolding', in *Senses and Society* 3(2): 187-204.

Lash, S. (2001), 'Technological forms of life', *Theory, Culture Society* 18: 105-120.

Lassen, C. (2006), 'Aeromobility and work', *Environment and Planning A* 38(2): 301-312.

Latham, A. and McCormack, D. (2004), 'Moving cities: rethinking the materialities of urban geographies', *Progress in Human Geography* 28(6): 701-724.

Latour, B. (1993), *We Have Never Been Modern* (Hemel Hempstead: Harvester Wheatsheaf).

Latour, B. (1997), 'Trains of thought: Piaget, formalism, and the fifth dimension', *Common Knowledge* 6(3): 170-191.

Latour, B. (2005), *Reassembling the Social: An Introduction to Actor-Network Theory* (Oxford: Oxford University Press).

Law, L. (2001), 'Home cooking: Filipino women and geographies of the senses in Hong Kong', *Ecumene* (8): 264-283.

Law, L. (2005), 'Sensing the City: Urban Experiences', in P. Cloke, P. Crang and M. Goodwin (eds), *Introducing Human Geographies*, 2nd edn (London: Arnold).

Lazzarato, M. (2004), 'From capital-labour to capital-life', *Ephemera* 4: 187-208.

Lee, J. (1989), *Ireland 1912-1985: Politics and Society* (Cambridge: Cambridge University Press).

Lefebvre, H. (1987), 'The everyday and everydayness', *Yale French Studies* 73: 7-11.

Lefebvre, H. (1991), *The Production of Space*, trans. by D. Nicholson-Smith (Oxford: Blackwell).

Lefebvre, H. (1996), *Writings on Cities* [ed. by E. Koffman and E. Lebas] (Oxford: Blackwell).

Lefebvre, H. (2004), *Rhythmanalysis: Space, Time and Everyday Life*, trans. S. Elden and G. Moore (London: Continuum).

Lefebvre, H. (2002), *Critique of Everyday Life Volume II: Foundations for a Sociology of the Everyday*, trans. John Moore (London: Verso).

Lefebvre, H. (2003), 'The other Parises', in S. Elden, E. Lebas and E. Koffman (eds), *Henri Lefebvre: Key Writings* (London: Continuum).

Lefebvre, H. (2008), *Critique of Everyday Life Volume 1*, trans. J. Moore (London: Verso).

Levinas, E. (2003), 'On the utility of insomnia (interview with Bertrand Revillon)', *Unforeseen History*, trans. N. Poller (Chicago: University of Illinois Press).

Mels, T. (2004), (ed.), *Reanimating Places: A Geography of Rhythms* (Aldershot: Ashgate).

Lindqvist, C. and Snickars, F. (eds) (1975), *Space-time and Human Conditions in Dynamic Allocation of Urban Space Vol. 3* (Westmead: Saxon House).

Lipsky, M. (1980), *Street-Level Bureaucracy* (New York: Russell Sage Foundation).

Liu, A. (2004), *The Laws of Cool: Knowledge Work and the Culture of Information* (Chicago: University of Chicago Press).

Longhurst, R. (2001), *Bodies: Exploring Fluid Boundaries* (London: Routledge).

Lorimer, H. (2005), 'Cultural geography: the busyness of being "more-than-representational"', *Progress in Human Geography* 29(1): 83-94.

Lorimer, H. (2007), 'Cultural geography: worldly shapes, differently arranged', *Progress in Human Geography* 31(1): 89-100.

Lynch, K. (1960), *The Image of the City* (Cambridge, MA: MIT Press).

Määttänen, P. (2005), 'Aesthetics of movement and everyday aesthetics', *Contemporary Aesthetics* 13 (Special Volume 1), available at: <http://www.contempaesthetics.org/newvolume/pages/journal.php?volume=13>, accessed 2 March 2009.

McCormack, D. (2002), 'A paper with and interest in rhythm', *Geoforum* 33(4): 469-485.

McCrossen, A. (2005), 'Sunday: marker of time, setting for memory', *Time and Society* 14(1): 25-38.

McCully, J. (2007), *Beyond the Moon: A Conversational, Common Sense Guide to Understanding the Tides* (Hackensack: World Scientific Publishing).

McLean, A. (2003), *The Truth about Horses: A Guide to Understanding and Training Your Horse* (Sydney: Barron's).

Macnaghten, P. and Urry, J. (1998), *Contested Natures* (London: Sage).

McKenzie, W. (1937), 'Belisha crossers can't be blamed', *Daily Mail* 20 July, 9.

McKenzie, W. (1939), 'Anti jay-walking squad go on duty', *Daily Mail* 3 July, 5.

Madson, P. (2005), *Improv Wisdom: Don't Prepare, Just Show Up* (New York: Bell Tower).

Madson, P. (2005a). *Improv Wisdom*, available at: <http://www.improvwisdom.com/>, accessed 20 November 2008.

Mann, M. (2004), 'Introducing the Hipster PDA', available at: <http://www.43folders.com/2004/09/03/introducing-the-hipster-pda>, accessed 20 April 2008.

Mann, M. (2005), 'Procrastination hack: "(10+2)*5"', Available at: <http://www.43folders.com/2005/10/11/procrastination-hack-1025>, accessed 4 March 2008.

Mann, M. (2008), '43 Folders: Time, Attention, and Creative Work', available at: <http://www.43folders.com/2008/09/10/time-attention-creative-work>, accessed 30 November 2008.

Manuel, P., Bilby, K. and Largey. M. (1995), *Caribbean Currents: Caribbean Music From Rumba to Reggae* (Philadelphia: Temple University Press).

Marx, K. (1973), *Grundrisse*, trans. M. Nicolaus (London: Penguin).

Massey, D. (1992), 'Politics and space/time', *New Left Review* 196: 65-84.

Massey, D. (1993), 'Power-geometry and a progressive sense of place', in J. Bird, B. Curtis, T. Putnam, G. Robertson and L. Tickner (eds), *Mapping the Futures: Local Cultures, Global Change* (London: Routledge).

Massey, D. and Thrift, N. (2003), 'The passion of place', in R. Johnston and M. Williams (eds), *A Century of British Geography* (Milton Keynes: Oxford University Press/British Academy).

Massey, D. (1994), *Space, Place and Gender* (Cambridge: Polity Press).

Massey, D. (1995), *Spatial Divisions of Labour: Social Structures and the Geography of Production*, 2nd Edition (London: Macmillan).

Massey, D. (1995), 'The conceptualization of place', in D. Massey and P. Jess (eds), *A Place in the World? Places, Cultures and Globalization* (London: Open University Press).

Massey, D. (2005), *For Space* (London: Sage).

Massey, D. (2007), *World City* (Cambridge: Polity).

Masters, A. (2006), *Stuart: A Life Backwards* (London: Harper Perennial).

Maurice, A. (2001), *House Doctor Quick Fixes: Top 100 Ways to Add Pounds, Style and Calm to Your Home* (London: HarperCollins).

Matless, D. (2009), 'Nature Voices', *Journal of Historical Geography* 35(1): 178-188.

May, J. and Thrift, N. (eds) (2001), 'Introduction' in J. May and N. Thrift (eds), *Timespace: Geographies of Temporality* (London: Routledge).

Mels, T. (2004), 'Lineages of a geography of rhythm', in T. Mels (ed.), *Reanimating Places: A Geography of Rhythms* (Aldershot: Ashgate).

Merker, B., Madison, G. and Eckerdale, P. (2009), 'On the role and origin of isochrony in human rhythmic entrainment', *Cortex* 451: 4-17.

Merrifield, A. (2006), *Henri Lefebvre: A Critical Introduction* (London: Routledge).

Merriman, P. (2005), 'Driving places: Marc Augé, non-places and the geographies of England's M1 motorway', in M. Featherstone, N. Thrift and J. Urry (eds), *Automobilities* (London: Sage).

Merriman, P. (2007), *Driving Sspaces: A Cultural-Historical Geography of England's M1 Motorway* (London: Blackwell), 514.

Meyer, K. (2008), 'Rhythms, Streets, Cities', in K. Goonewardena et al. (eds), *Space, Difference, Everyday Life* (New York: Routledge).

Minister of Transport (1934), *The London Traffic (Pedestrian Crossing Places), Provisional Regulations* (London: HMSO), 5.

Molotch, H., Freudenburg, W. and Paulsen, K. (2000), 'History repeats itself, but how? city character, urban tradition and the accomplishment of place', *American Sociological Review* 65(6): 791-823.

Moore, R. (2006), *Music and Revolution: Cultural Change in Socialist Cuba* (Berkeley: University of California Press).

Moores, S. (1995), 'TV discourse and "time-space distanciation": on mediated interaction in modern society', in *Time and Society* 4(3): 329-44.

Moran, J. (2004), 'History, memory and the everyday', *Rethinking History* 8(1): 51-68.

Morin, K. (1999), 'Peak practices: Englishwomen's "heroic" adventures in the nineteenth-century American West', *Annals of the Association of American Geographers* 89(3): 489.

Morton, T. (2007), *An Ecology without Nature: Rethinking Environmental Aesthetics* (Cambridge, MA: Harvard University Press).

Murdoch, J. (1998), 'The spaces of actor-network theory', *Geoforum* 29(4): 357-374.

Nash, C. (2000), 'Performativity in practice: some recent work in cultural geography', *Progress in Human Geography* 24(4): 653-664.

Nash, C. (2005), 'Geographies of relatedness', *Transactions of the Institute of British Geographers* 30(4): 449-462.

Nature Conservancy Council (1991), *Nature Conservation and Estuaries in Great Britain* (Peterborough, NCC).

Newman J. (n.d.), 'What is Dressage?', *Equiworld: the Internet's most extensive equestrian resource* <http://www.equiworld.net/uk/sports/dressage/dressage.htm>.

Nichols, R. (1957), *Wings for Life* (Philadelphia and New York: J.B. Lippincott).

Normark, D. (2006), 'Tending to mobility: intensities of staying at the petrol station', *Environment and Planning A* 38(2): 241-252.

Norton, P. (2007), 'Street rivals: jaywalking and the invention of the motor age street', *Technology and Culture* 28(2): 331-359.

Noteberg, S. (2008), 'Programming with underwear inside out', available at: <http://blog.staffannoteberg.com/2008/01/31/programming-with-underwear-inside-out/>, accessed 12 December 2008.

Novak, J. and Sykora, L. (2007), 'A city in motion: time-space activity and mobility patterns of suburban inhabitants and the structuration of the spatial organisation of the Prague metropolitan area', *Geografiska Annaler* 89(2): 147-167.

O'Brien, D. (2004), 'Life Hacks: Tech Secrets of Overprolific Alpha Geeks. Notes', available at <http://craphound.com/lifehacksetcon04.txt>, accessed 4 March 2008.

O'Connell, M. (2001), *Changed Utterly: Ireland and the New Irish Psyche* (Dublin: Liffey Press).

O'Gorman, M. (1929), 'Road accidents', *The Times*, 13 September, 13.

O'Neill, J. (2007), *Markets, Deliberation and Environment* (Abingdon: Routledge).

O'Reilly, K. (2000), *The British on the Costa del Sol: Transnational Identities and Local Communities* (London: Routledge).

Orovio, H. (1985), 'La Rumba del Tiempo de España', *Revolución y Cultura*, December: 56-61.

Packer, J. (2008), *Mobility Without Mayhem: Safety, Cars and Citizenship* (Durham: Duke University Press).

Parkes, D. and Thrift, N. (1978), 'Putting time in its place', in T. Carlstein, D. Parkes and N. Thrift (eds), *Making Sense of Time* (London: Edward Arnold).

Parkin, J., Ryley, T. and Jones, T. (2007), 'Barriers to cycling: an exploration of quantitative analyses', in D. Horton, P. Rosen and P. Cox (eds), *Cycling and Society* (Aldershot: Ashgate).

Parkins, W. (2004), 'Out of time: fast subjects and slow living', in *Time and Society* 13(2/3): 363-382.

Pascoe, P. (2001), *Airspaces* (London: Reaktion).

Pearson, K. (2002), *Philosophy and the Adventure of the Virtual* (London: Routledge).

Pérez, L. (2008), *On Becoming Cuban: Identity, Nationality and Culture*, 2nd edn, (Chapel Hill: University of North Carolina Press).

Phelan, P. (1996), 'Dance and the history of hysteria', in S. Foster (ed.), *Corporealities: Dancing Knowledge, Culture and Power* (London: Routledge).

Pine, B. and J. Gilmore (1999), *The Experience Economy* (Boston: Harvard Business School Press).

Pink, S. (2007), 'Sensing cittàslow: slow living and the constitution of the sensory city', in *Senses and Society* 2(1): 59-78.

Plath, S. (1981), *Collected Poems* (London: Faber & Faber).

PocketMod (2007), 'The PocketMod: The Free Recyclable Personal Organiser', Available at: <http://www.pocketmod.com/>, accessed 20 April 2008.

Potts, T. (2008), 'Orchestrating chaos: clutter, rhythm and everyday life', *Keywords: A Journal of Cultural Materialism* 5: 88-105.

Pred, A. (1977), 'The choreography of existence: comments on Hagerstrand's time-geography and its usefulness', *Economic Geography* 53(2): 207-221.

Price, J. and Shildrick, M. (1999), 'Openings on the body: a critical introduction', in J. Price and M. Shildrick (eds), *Feminist Theory and The Body* (London: Routledge).

Procalidad (2002), *Indice Nacional de Satisfaccion de Consumidores, 1er semester, 2002* (Santiago: Procalidad).

Punter, J. (2005), 'City centre and bay', in M. Ungersma (ed.), *Cardiff: Rebirth of a Capital* (Cardiff: Cardiff Council).

Punter, J. (2006), 'A city centre for a European capital?', in A. Hooper and J. Punter (eds), *Capital Cardiff 1975-2020* (Cardiff: University of Wales Press).

Raban, J. (1992), *The Oxford Book of the Sea* (Oxford: Oxford University Press).

Revill, G. (2004), 'Performing French folk music: dance, authenticity and non-representational theory', *Cultural Geographies* 11: 199-209.

Richardson, P. (2004), 'West Side Story', *Conde Nast Traveller*, June.

Roberts, F. (1933), 'Legs', *The Times*, 7 August, 11.

Robinson, Bruce, dir. (1987), *Withnail and I* (UK: Handmade Films).

Robinson, C. (2007), *Between the Tides, The Perilous Beauty of Morecambe Bay* (Ilkley: Great Northern Books).

Roche, M. (2003), 'Mega-events, time and modernity: on time structures in global society', in *Time and Society* 12(1): 99-126.

Rodriguez, R. (1977), 'La rumba en la provincia de Matanzas', *Boletín de Música* (65): 15-23.

Rowe, D. (2004), 'The assessment of risk is a very personal affair', in *CABE What Are We Scared Of? The Value of Risk in Designing Public Space* (London: CABE Space).

Rowe, M. (1999), *Crossing the Border: Encounters Between Homeless People and Outreach Workers* (Berkeley: University of California Press).

Rybczynski, W. (1999), *A Clearing in the Distance: Frederick Law Olmsted and North America in the Nineteenth Century* (Toronto, CA: HarperCollins).

Sabbagh, J. (2006), 'Estaciones de servicio', *Revista ARQ* 62: 62-65.

Sacks, O. (2007), *Musicophilia: Tales of Music and the Brain* (London: Picador).

Sagoff, M. (1988), *The Economy of the Earth* (Cambridge: Cambridge University Press).

Sandhu, S. (2006), *Night Haunts: A Journey Through London's Night* (London: Verso).

Sarduy, P. (2007), 'An open letter to Carlos Moore', AfroCubaWeb, 24 February 2007, available at <http://afrocubaweb.com/lettertocarlos.htm>, accessed 6 December 2007.

Savage, M., Bagnall, G. et al. (2005), *Globalisation and Belonging* (London: Sage).

Schaffer, M. (1977), *The Tuning of the World* (New York: Knopf).

Schwanen, T. (2007), 'Matter(s), of interest: artefacts, spacing and timing', *Geografiska Annaler* 89B (1): 9-22.

Seamon, D. (1980), 'Body-subject, time-space routines, and place-ballets', in A. Buttimer and D. Seamon (eds), *The Human Experience of Space and Place* (London: Croom Helm).

Seigworth, G. and Gardiner, M. (2004), 'Rethinking everyday life', *Cultural Studies* 18(2): 139-159.

Sennett, R. (1994), *Flesh and Stone: The Body and the City in Western Civilization* (London: Faber & Faber).

Sennett, R. (2004), *Respect: The Formation of Character in a World of Inequality* (London: Penguin).

Severn Estuary Partnership (2005), *Strategy for the Severn Estuary* (Cardiff: SEP).

Shackley, S. and Wynne, B. (1996), 'Representing uncertainty in global climate change science and policy: boundary-ordering devices and authority', *Science, Technology and Human Values* 21(3): 275-302.

Shapiro, N. and Hentoff, N. (1955), *Hear Me Talkin' to Ya: The Story of Jazz as Told by the Men Who Made It* (New York: Dover).

Shaw, J. (2001), '"Winning territory": changing place to change place', in J. May and N. Thrift (eds), *Timespace: Geographies of Temporality* (London: Routledge).

Sheller, M. (2004), 'Automotive emotions: feeling the car', *Theory, Culture & Society* 21(4/5): 221-242.

Sheller, M. and Urry, J. (2006), 'The new mobilities paradigm', *Environment and Planning A: Environment and Planning* 38: 207-226.

Silverman, H. (1975), 'Review of *The Phenomenology of Aesthetic Experience* by Mikel Dufrenne', *The Journal of Aesthetics and Art Criticism* 33 (4): 462-464.

Silverstone, R. (1994), *Television and Everyday Life* (London: Routledge).

Simmel, G. (1969), 'The metropolis and mental life', in R. Sennett (ed.), *Classic Essays on the Culture of Cities* (New York: Appleton-Century-Crofts).

Simpson, P. (2008), 'Chronic everyday life: rhythmanalysing street performance', *Social and Cultural Geography* 9(7): 807-829.

Smith, A. (2005), 'Conceptualizing city image change: the 're-imaging' of Barcelona', *Tourism Geographies* 7(4): 398-423.

Southerton, D. (2003), '"Squeezing time": allocating practices, coordinating networks and scheduling society', *Time and Society* 12(1): 5-25.

Sparks, C. (2006), 'A stranger abroad in Leitrim', in M. Corcoran and M. Peillon (eds), *Uncertain Ireland: A Sociological Chronicle 2003-2004* (Dublin: Institute of Public Administration).

Spinney, J. (2006), 'A place of sense: A kinaesthetic ethnography of cyclists on Mt Ventoux', *Environment and Planning D: Society & Space* 24(5): 709-732.

Spinney, J. (2007), 'Cycling the city: non-place and the sensory construction of meaning in a mobile practice', in D. Horton, P. Rosen and P. Cox (eds), *Cycling and Society* (Aldershot: Ashgate).

Spinney, J. (2009), 'Cycling the city: movement, meaning and method', *Geography Compass* 4: February.

Solnit, R. (2001), *Wanderlust: A History of Walking* (London: Verso).

Stern, N. (2006), *Stern Review on the Economics of Climate Change* (London: New Economics Foundation).

Subirats, J. and Rius, J. (2005), *Del Xino al Raval: Cultura Itransformacio Social a la Barcelona Central* (Barcelona: Centro de Cultura Contemporaneo de Barcelona).

Sublette, N. (2004), *Cuba and Its Music: From the First Drums to the Mambo* (Chicago: Chicago Review Press).

Swyngedouw, E. (2007), 'Impossible sustainability and the post-political condition', in R. Krueger and D. Gibbs (eds), *The Sustainable Development Paradox: Urban Political Economy in the US and Europe* (New York: Guilford Press).

Szerszynski, B. and Urry, J. (2006), 'Visuality, mobility and the cosmopolitan: inhabiting the world from afar', *The British Journal of Sociology* 57(1): 113-131.

Taylor, F.W. (1998/1911), *The Principles of Scientific Management* (Mineola, NY: Dover).

Thaden, L. [1938] (2004), *High, Wide, and Frightened* (Little Rock, AR: University of Arkansas Press).

Thomas, H. (2003), *Discovering Cities: Cardiff* (Sheffield: Geographical Association).

Thrift, N. (1997), 'The still point: resistance, expressive embodiment and dance', in S. Pile and M. Keith (eds), *Geographies of Resistance* (London: Routledge).

Thrift, N. (1999), Steps to an ecology of place', in D. Massey, P. Sarre, and J. Allen (eds), *Human Geography Today* (Oxford: Polity).

Thrift, N. (2004), 'Driving in the city', *Theory, Culture & Society* 21(4/5): 41-59.

Thrift, N. (2004), 'Movement-space: the changing domain of thinking resulting from the development of new kinds of spatial awareness', *Economy and Society* 33(4): 582-604.

Thrift, N. (2005), 'But malice aforethought: cities and the natural history of hatred', *Transactions of the Institute of British Geographers* 30: 133-150.

Times, The (1927), 'Legs and wheels', 2 April, 3.

Times, The (1927), 'The third condition', 22 April, 13.

Times, The (1938), 'Six Miles of Guard Rails', 29 May, 11.

Timmermans, H., van der Waerden, P., Alves, M., Polak, J., Ellis, S., Harvey, A., Kurose, S. and Zandee, R. (2003), 'Spatial context and the complexity of daily travel patterns: an international comparison', *Journal of Transport Geography* 11: 37-46.

Tomlinson, J. (2007), *The Culture of Speed: The Coming of Immediacy* (London: Sage).

Tripp, H. (1938), *Road Traffic and Its Control* (London: Edward Arnold).

Turing, A. (1937), 'On computational numbers, with an application to the *Entscheidungsproblem*', *Proceedings of the London Mathematical Society* 2(42): 230-265.

Turing, A. (1951), 'Computing machinery and intelligence', *Mind* 59(236): 433-460.

Turner, G. (1964), *The Car Makers* (Harmondsworth: Penguin).

Ungersma, M. (ed.), (2005), *Cardiff: Rebirth of a Capital* (Cardiff: Cardiff Council).

Urry, J. (1990), *The Tourist Gaze* (London: Sage).

Urry, J. (2000), *Sociology Beyond Societies: Mobilities for the Twenty-First Century* (London: Routledge).

Urry, J. (2002), 'Mobility and proximity', *Sociology* 36(2): 255-274.

Urry, J. (2003), 'Social networks, travel and talk', *British Journal of Sociology* 54(2): 155-175.

Urry, J. (2003), Workshop 1: 'Beyond societies', International Colloquium, *The Sense of Movement: Modernity and Mobilities in Contemporary Urban Societies*, organised by IVM, Centre culturel international de Cerisy-la-Salle, Paris, France, Institute pour la ville en mouvement.

Urry, J. (2006), 'Travelling Times', *European Journal of Communication* 21(3): 357-372.

Virilio, P. (2007), *The Original Accident* (Cambridge: Polity Press).

von Holst, E. (1973), *The Behavioral Physiology of Man and Animals* The Collected Papers of Erich von Holst (Coral Gables, FL: University of Miami Press).

Vowles, T. (2006), 'Geographic perspectives on air transportation', *Professional Geographer* 58(1): 12-19.

Wade, P. (2002), *Race, Nature and Culture: An Anthropological Perspective* (London and Sterling, Virginia: Pluto Press).

Wade, P. (2004), 'Human nature and race', *Anthropological Theory* 4(2): 157-172.

Wade, P. (2005), 'Rethinking Mestizaje: ideology and lived experience', *Journal of Latin American Studies* 37: 239-257.

Wacjman, J. (2008), 'Life in the fast lane? Towards a sociology of technology and time', *British Journal of Sociology* 59(1): 59-77.

Walter, B. (2001), *Outsiders Inside: Whiteness, Place and Irish Women* (London: Routledge).

Walter, D. and Franks, M. (2002), *The Life Laundry: How To De-Junk Your Life* (London: BBC).

Walter, F. (2008), *Catastrophes: une histoire culturelle: XVIe-XXIe siècle* (Paris: Seuil).

Watson, L. (1973), *Supernature: The Natural History of the Supernatural* (London: Book Club Associates).

Watts, L. and Urry, J. (2008), 'Moving methods, travelling times', *Society and Space: Environment and Planning D* 26(5): 860-874.

Wells, M. (2000), 'Office clutter or meaningful personal displays: the role of office personalization in employee and organizational well-being', *Journal of Environmental Psychology* 20(3): 239-255.

Whatmore, S. (1999), 'Rethinking the "human" in human geography' in D. Massey, P. Sarre and W. Wheeler, *A New Modernity? Change in Science, Literature and Politics* (London: Lawrence and Wishart).

Whatmore, S. (2006), 'Materialist returns: practising cultural geography in and for a more-than-human world', *Cultural Geographies* 13: 600-609.

Wiener, N. (1950), *The Human Use of Human Beings: Cybernetics and Society* (New York: Da Capo).

Wikipedia, 'Life hack', available at: <http://en.wikipedia.org/wiki/Life_hack>, accessed 8 November 2008.

Williams, R. (1961), *The Long Revolution* (London: Chatto and Windus).

Winkler, J. (2002), 'From acoustic horizons to tonalities', in J. Winkler (ed.), *Space, Sound and Time: A Choice of Articles in Soundscape Studies and Aesthetics of Environment 1990–2003*, available at <http://www.humgeo. unibas.ch/homepages/anhaenge_hompages/winkler_space-sound-time_08-08-12ii.pdf>, accessed 2 March 2009.

Winter, J. (1993), *London's Teeming Streets, 1830-1914* (London: Routledge).

Wrightson, K. (2000), 'An introduction to acoustic ecology', *The Journal of Acoustic Ecology* 1(1): 10-13.

Wunderlich, F. (2007), 'Walking and rhythmicity: sensing urban space', in *Journal of Urban Design* 13(1): 31-44.

Wylie, J. (2005), 'A single day's walking: narrating self and landscape on the South West Coast Path', *Transactions of the Institute of British Geographers* 30(2): 234-247.

Yeats, W.B. [1921] (1970), *Michael Robartes and the Dancer* (Shannon: Irish University Press).

Yelanjian, M. (1991), 'Rhythms of consumption', *Cultural Studies* 5(1): 91-97.

Young, I. (1990), 'Throwing like a girl: a phenomenology of feminine body comportment, motility, and spatiality', in I. Young, *Throwing Like a Girl and Other Essays in Feminist Philosophy and Social Theory* (Bloomington, IN: Indiana University Press).

Young, I. (1997), *Intersecting Voices: Dilemmas of Gender, Political Philosophy and Policy* (Princeton, NJ: Princeton University Press).

Young, M. (1988), *The Metronomic Society: Natural Rhythms and Human Timetables* (London: Thames and Hudson).

Zerubavel, E. (1981), *Hidden Rhythms: Schedules and Calendars in Social Life* (Chicago and London: University of Chicago Press).

Žižek, S. (2008a), 'Interview with Amy Goodman', *Democracy Now!* 12 May 2008, available at: <http://www.democracynow.org/2008/5/12/world_renowned_ philosopher_slavoj_zizek_on>, accessed 19 January 2009.

Žižek, S. (2008b), *In Defence of Lost Causes* (London: Verso).

Zukin, S. (1995), *The Cultures of Cities* (Oxford: Blackwell).

Zukin, S. (2008), 'Consuming Authenticity', *Cultural Studies* 22(5): 724-748.

Index